What does the science of animal intelligence mean for how we understand and live with the wild creatures around us?

Honeybees deliberate democratically. Rats reflect on the past. Snakes have friends. In recent decades, our understanding of animal intelligence has exploded—but the way we relate to wild animals has yet to catch up. *Meet the Neighbors* asks: What would it mean if we took animal minds seriously? Science shows that the creatures who share our everyday landscapes—from devoted geese to clever raccoons to playful turtles—are thinking, feeling individuals. Should we then see them as fellow persons, even citizens? Weaving in the latest research, Brandon Keim takes us into courtrooms and wildlife hospitals, under backyard decks and into deserts, to meet the philosophers, rogue pest controllers, ecologists, wildlife doctors, and others who are reimagining our relationships to animals and to nature. A beguiling invitation to discover an expanded sense of community and kinship beyond our own species, *Meet the Neighbors* opens our eyes to the world of vibrant intelligence just outside our doors.

Brandon Keim

Brandon Keim is an independent journalist specializing in animals, nature, and science. His work appears regularly in the *New York Times*, the *Atlantic*, *Nautilus*, *National Geographic*, and elsewhere. He lives in Bangor, Maine.

Meet the Neighbors

ANIMAL MINDS AND LIFE IN
A MORE-THAN-HUMAN
WORLD

Meet the Neighbors

Brandon Keim

Illustrations by
MATTIAS LANAS

W. W. NORTON & COMPANY
Independent Publishers Since 1923

For Aurora,
who made this book possible.

Even though I usually don't
know this squirrel from another,
I know that if I tried,
I would....

—BARBARA SMUTS,
"ENCOUNTERS WITH ANIMAL MINDS"

Contents

IV. IN THE WILD

Meet the
Neighbors

Introduction

ONE BLUSTERY, SLEETING TUESDAY MORNING IN APRIL, while preparing her garden for spring, Fiona Presly spotted a bumblebee. She had just finished covering up several newts whom she accidentally disturbed while moving a bag of compost; when she looked up, there was the bee, bedraggled in the cold wet grass beside her.

Presly has a particular fondness for bumblebees. When she was a little girl, her father taught her how to stroke them: slowly and gently, attentive to the raised-middle-leg gesture that signals bumblebee irritation. If she sees one on a footpath—as sometimes happens when they tire, or near their lives' end—she moves the bee safely to one side. Presly put her hand, warm from exertion, in front of the bee, who promptly crawled into her palm.

Taking a closer look, Presly noticed two things. The pattern of stripes told her that she had met a queen *Bombus terrestris*, or buff-tailed bumblebee. This queen also had no wings. A virus had likely deformed them, although Presly wouldn't learn about that until later. First she took the wingless queen inside, mixed a spoonful of sugar water, watched as the bee took a sip and primly cleaned her fur, and considered.

"I thought, well, what can I do? I have to go to work," recalls Presly, a library assistant in the coastal village of Culloden, Scotland, describing a dilemma familiar to anyone who has stopped to help an animal in distress without worrying about the practicalities. She put the bee back in her garden and went in for her shift. When she returned, the bee was in the same place, looking worse for the wear. Presly took her inside and gave her more sugary water. Again the bee crawled into her hand.

For the next few weeks Bee—as Presly named her—lived in a box that Presly set up in the kitchen. On sunny days she would take Bee outside. By May it was warm enough for her to be outdoors full-time; she planted pots of scabious and heather and anemones from the local garden center and enclosed them in mesh, but with gaps so that Bee could leave if she wanted. There came three anxious days when Presly didn't see Bee. On the fourth morning, as she was feeding a neighbor's cat, Presly's husband texted: Bee had returned! She dashed home. There was Bee, disheveled but alive, ready to be picked up.

A golden summer passed this way. Bee lived in her garden; Presly would visit and, if she didn't see Bee right away, tapped lightly on her nest box. Out Bee would come, antennae high. Sometimes she liked to perch on Presly's nose, but most of the time she stayed in Presly's hand. Presly wonders now whether her warmth and the steady rhythm of her pulse reminded Bee of a colony's humming bustle.

"She liked to sit in the crook of my hand and clean. Then her antennae would go right down and she would have a sleep. I'd sit and read a book," says Presly. Once, when Presly had taken her inside on a stormy day, she found Bee with her back turned and, so as not to scare her, announced her presence by scratching the table. Bee turned around and trotted right to her. "Cross my heart, I would say that Bee was always pleased to see me," Presly remembers. "She wasn't looking for anything to eat or drink. She would just come and snuggle on my hand. It's almost as if the company meant as much to her as it meant to me."

Every so often Bee would try to fly, vibrating her wing muscles fruitlessly. Afterward she drooped and hid her face. Presly would stroke her and murmur; Bee would click and hum and start to clean herself. Bee was fastidious: she didn't defecate in her nest and, if she needed to relieve herself while in Presly's hand, would poke her abdomen off the side. Presly liked to joke that Bee was potty-trained.

As summer progressed Bee stopped trying to fly. She was no longer young; indeed she was already middle-aged when Presly met her, having been born the previous summer and hibernating through fall and winter. If a dog's year is seven of ours, then so is a bee's month. By late summer her time was nearly up. She slept more and more. She continued to drink but stopped eating pollen. Her movements slowed.

"Did I cry? Yes," Presly tells me with a small, complicated laugh. She knew that some people would consider this ridiculous, but it mattered a great deal to her. "She was in my hand when her last wee leg moved. We buried her in the garden. Nobody would know where she is, but I know where she is, beside one of her favorite heathers." As we talk, Presly's voice still catches. Bee might only have been a bee, but she was a friend. "People might think I was imagining it," Presly continues, yet she believed that Bee knew and recognized her. "I can still feel her in my hands," Presly says. "She was quite incredible, really."

There's a chance that you already know of Presly and Bee. I learned about them from a video on *The Dodo*, a website that specializes in aww-inducing stories about animals. When Presly last spoke to me the video had been viewed 124 million times—a remarkable number even by the standards of its genre, of which I am an unapologetic enthusiast. Western civilization in the early twenty-first century is fractured in many ways, but we can still be united by a skyscraper-climbing raccoon or a goose looking through the door of an animal hospital where her mate is being treated for a broken foot.

Now, such stories may strike some readers as clickbait distractions from grim reality. Humanity is obliterating species at rates last seen when an asteroid struck Earth, sending up a shroud of dust and smoke that blotted out the sun; most of those not going extinct are dwindling. In the face of this cataclysm, this planetary ecocide, we have people building prosthetic wings for parrots and making friends with octopuses.

Yet there's more to these stories than low-hanging sentiment. The world's problems are so vast and painful as to be paralyzing; narratives of connection are sparks of hope, of unvanquished possibility. These stories are also rooted in empathy: in a sense of other animals as thinking, feeling beings with whom, despite all our differences, we have much in common.

That is a lesson of contemporary science, whose journals now swell with insights into animal minds. Even Bee's tiny brain was remarkably complex. Scientists have described how bumblebees learn by observation and by association; count; recognize by sight what they've previously only touched, and vice versa; are capable of using tools; and have feelings. Lars Chittka, an entomologist who studies bee cognition, wrote in a commentary on Bee that—as Presly intuited—she quite likely learned to recognize Presly by scent, and then to associate Presly with food and warmth. With time that relationship transmuted into something that, were Bee a dog or cat, would readily be seen as affection. Though Bee last shared a common ancestor with *Homo sapiens* 600 million years ago and was even tinier in comparison to Presly than Presly would be to the world's largest tree, she was still someone you could relate to.

◆ ◆ ◆

HARDLY A WEEK PASSES WITHOUT STORIES OF SOME delightful new scientific insight into the minds of animals: snake friendships and songbird syntax, bison deliberations and puzzle-solving turtles, self-aware fish and rats who feel regret. Chittka calls it a "new Copernican revolution," as profound as the realiza-tion that Earth revolves around the sun, rather than vice versa. Humans are merely one star in the cosmos of animal intelligence.

Yet despite what scientists have learned, it often seems as if their findings get stuck in translation. They don't enter our minds once we shut the screen and go outside. Even as we recognize our beloved pets as thinking, feeling beings with a first-person expe-rience of life, and grapple—however inconsistently—with the self-hood of animals used for food and research, that's not how we're socialized to regard wild animals.

Instead the institutions and concepts that frame our relations to nature are rooted in traditions that denied nonhuman intelligence. Beauty and wonder, transcendence, resources, wildlife manage-ment, stewardship, conservation, biodiversity, environmentalism, sustainability, even natural history: animals as meaningfully intel-ligent beings, as individuals who share essential mental properties with us, only rarely appear. And though people doing the work may think otherwise, that sensibility has been relegated to children's sto-ries and half-serious entertainments—important genres, but easily dismissed. Our cultural habits of mind, when it comes to nature and the animals living outside our homes, are a legacy of centuries past.

The science of animal intelligence challenges those habits. In *Meet the Neighbors*, we'll begin by taking the latest research out into the everyday landscape of a suburban neighborhood. This is not a book only about the suburbs—but that's a good place to consider the world of minds around us, to bridge abstracted and lived appreciation. So often celebrations of animal intelligence focus on a select few stars, such as chimpanzees and dolphins, but there are worlds of thought and feeling and relationships all around us.

Simply being aware that humans and other animals are not categorically separated by some clear, bright line is only the beginning. With a sense of kinship comes the question of what we do with it. How might an awareness of animal minds shape the ways we understand them and, ultimately, how we live with them on this shared, precious planet? That is a central question of our time, and it animates the rest of this book.

Fiona Presly befriended a bumblebee, and part of the beauty of their story is that Presly seems emblematic. She is not a dyed-in-the-wool conservationist; she's simply someone who loves nature, who was taught by her parents to notice wildflowers and look for frog eggs, who thrills at an osprey circling overhead and leaves part of her lawn unmowed for the butterflies. Neither is Presly is an animal rights campaigner; she has a vegan colleague at the library whose ideas she has come to appreciate, and she believes that all living beings deserve respect, but she's still trying to work out what that means for herself. I think many of us can relate.

We'll meet other people who have taken these ideas further, fighting to have animals recognized as legal persons—a designation long reserved for humans alone—and even to give them voice in political systems. Then we'll explore how new relations to wild animals are taking shape in everyday life. It's easy to be broad-minded when animals are seen at a distance, being pretty; but what do we owe so-called pests, or animals who are sick or injured? How do we live with predators whose presence is not always welcome?

Those stories are set largely in cities and suburbs, the environs where most of us come to know nature, amid the creatures who populate our daily lives. From there we travel to wilder places, reckoning with what a deeper appreciation of animal minds—and the ethical questions that come with it—means for conservation and ecology. If wild animals are kin, how should they be managed? And might thinking this way lead to new conceptions of nature?

The answers to these questions are central to how we understand ourselves and the rest of the living world. And although thinking of other animals as fellow persons won't automatically halt mass extinction and replenish Earth's biodiversity, such a change of heart seems a prerequisite to avoiding those tragedies, just as the problems of human societies can only be solved if people care about one another.

By now you're likely familiar with the term "Anthropocene," which originated early this millennium to describe the magnitude of human influence on Earth's life and processes. Humanity has become a planetary force, regulating global nutrient cycles and changing the climate and rearranging ecosystems. This power is something to understand not only in terms of what we do, but in terms of who we are. Our values are inscribed on the planet. What and who we care about, how we see ourselves and others, our morals and our ethics: these now shape the course of life itself. And even if we prove unable to halt the ecological collapse, if our grandchildren's Earth is a radically different and less verdant place than our own, there will still be nature, and animals, and the imperative to live well with them.

The environmentalist Stewart Brand famously said, "We are as gods and might as well get good at it." For a long time that statement grated on me. It seemed arrogant, even triumphal, uncritically accepting human domination of Earth rather than stepping humbly back to be, as the pioneering conservationist Aldo Leopold put it, a plain member and citizen of the natural world.

Yet there is a truth to Brand's words. As individuals, each of us has the power to make decisions that transform, for better or worse, the lives of other creatures; and collectively, as communities, societies, and ultimately a civilization of eight billion human beings, our power is shaping the world. Whether or not we like it, we are indeed as gods. But we could turn the phrase a bit differently than Brand: We might as well be good neighbors.

I

A NEIGHBORHOOD
OF MINDS

1

The Essence of Intelligence

ON MOST MORNINGS THE LOCAL BIRDS WOULD GATHER in a little stream that fed the neighborhood stormwater retention pond. There, below the drainage pipe where it emerged, more a trickle than a stream, really, under the snakeweed and oriental bittersweet vines, they would come to bathe: house sparrows and song sparrows and starlings, boisterous mockingbirds and shy catbirds, cardinals, robins, sometimes a blue jay flashing iridescent.

They would sit in the shallows, splashing their wings, on a bed of gravel that sparkled when it was sunny. Sometimes they bathed alone but often in pairs or small groups. Watching them while I drank my coffee and planned the day became part of my morning ritual, and some months after I first noticed them, I considered that it was their morning ritual, too. They were like locals at a coffee shop or the gym. They were my *neighbors*.

This was the lesson of the third sort of nature I've known in my life. The first, around the town of Bangor, Maine, where I grew up, nourished a great-outdoors sort of appreciation: I couldn't identify more than handful of species or explain how nutrients cycled between stream and forest, but I learned the pleasure of remoteness, the feeling of pine needles, the beauty of a loon's call and a Milky Way undimmed by streetlights.

Later came New York City, where I lived for a decade across my twenties and thirties, during which time I became a science and nature writer. There I started to learn details—names of flowers in vacant lots, histories of migrating shorebirds who stopped in wetlands at the city's edges—and ecology, and to examine the ideals of wilderness and wildness that framed my relationship to nature. I also wrote about the science of animal intelligence and the ethics of how we treat nonhumans. The question arose, if not the answer, of what it could mean to think of wild animals as the thinking, feeling individuals they are.

Next came Bethesda, Maryland, a suburban city outside Washington, DC, where my girlfriend had a job, home to the stormwater retention pond. It took nearly a year to wrap my head around that environment. It didn't fit the frames of rural abundance or sidewalk-crack resilience to which I'd been accustomed. It was both dominated by humans and overflowing with animal life.

Once I learned to appreciate that, the sense of neighborliness followed—which may not sound revelatory, but it constituted a profound shift. Animals became individuals rather than, like the

drawings in a field guide, examples of their species. Never mind that I couldn't identify each sparrow individually; I knew *they* could. I stopped thinking of those animals as interlopers in an essentially human space. Instead they were residents of a place shared by people of many species, only some of them human.

This way of thinking is one implication of the flourishing science of animal intelligence. Spend much time with that research and it's soon apparent that what supposedly 'makes us human' is not, as is usually implied by that phrase, what makes us different from other creatures. It's what we have in common.

Truly we are kin. That idea is certainly not restricted to suburbs, of course; it's relevant just about anywhere humans or our impacts are found, which now means most everywhere on Earth. But the suburbs are an excellent place to pick up on it, so let's begin our journey with a walking tour of my old Bethesda neighborhood and the animal minds found there. You probably can relate to the setting, in its outlines if not in its species-by-species particulars. Wherever you are, whatever biome or urban configuration, you will have nonhuman neighbors of your own.

Out the apartment-building door we go one fine spring morning. To the left, in the same building, is a Mexican restaurant that stores its garbage in ill-secured dumpsters that fed a rat family until someone paved over the little patch of dirt where they'd dug homes. We'll return to rats—those exceptionally smart, persecuted creatures—in a little while. First, though, look across the street, to the low brick apartment complex beside the parking garage. There on a halfhearted strip of turf grass and big box-store shrubs a robin is hunting for earthworms.

It's the robin who naturally captures your gaze. *Turdus migratorius*, a member of the thrush family, more comfortable amid the urban bustle than his retiring forest relatives. His rich persimmon breast is a happy reminder that beauty can be quite common. Were Charles Darwin at your side, though, his attention might have focused on the earthworm.

Late in Darwin's life, when he'd already authored *The Origin of Species* and the lesser-known but comparably significant *The Expression of Emotions in Man and Animals*, Darwin spent several years—that's right, *years*—studying earthworms. His final book, *The Formation of Vegetable Mould Through the Action of Worms, with Observations on Their Habits*, was published a few months before his death. It became a surprise best-seller. Through Darwin's eyes, a despised pest was recast as an industrious continent-shaper whose tunnels leavened soil that would become impoverished without them.

Darwin also devoted much of the book to observations of earthworm behavior. He cataloged the types and shapes of leaf fragments that earthworms use to plug the openings of their burrows; found their arrangement to be ordered rather than random; and designed experiments to determine whether this was purely instinctive or a matter of choice. It appeared, he concluded, to be the latter. "If worms have the power of acquiring some notion, however rude, of the shape of an object and of their burrows," wrote Darwin, "they deserve to be called intelligent; for they then act in nearly the same manner as would a man under similar circumstances."

He acknowledged their awareness, their *inner lives*, the subjectivity implicit in an earthworm's preference for a warm, cozy burrow. Although Darwin didn't present his findings as conclusive, and the jury is still out on exactly what earthworms think while burrowing, the very fact of his open-mindedness to the possibilities of their minds was a radical thing.

For more than two thousand years, belief in animal *un*intelligence had spread through western civilization. Aristotle, zoologist as well as philosopher, thought animals capable of feeling only pain or hunger; his hierarchical taxonomy of life, with humans perched on top, was adapted by Christian theologians bent on replacing pagan beliefs in human–animal kinship. They emphasized rationality as a supposedly defining human trait and laid the foundations for Enlightenment thinkers such as Nicholas Malebranche, a seventeenth-century philosopher who neatly distilled the attitude of his peers when he said animals "eat without pleasure, cry without pain, grow without knowing it; they desire nothing, fear nothing, know nothing."

Not everyone agreed, certainly, but for the next few centuries, most serious-minded people did. Darwin refuted that convention. *The Expression of Emotions* taught that evolution didn't furnish humans and other animals only with common anatomical traits, but with mutual properties of mind as well. His earthworms were an exclamation point.

After Darwin passed, open-mindedness to animal intelligence became respectable in scientific circles. It didn't last long, though, and for most of the twentieth century, mainstream scientific convention again treated animals as essentially mindless. We'll return to that history later. For the purposes of our walk through the neighborhood, it's enough simply to know that research on animal minds has again flourished. Now it's time to take those insights out of scientific journals and into our everyday world.

◆　◆　◆

WALK DOWN THE BLOCK A LITTLE WAY, PAST A CON-venience store, a salon, a supermarket, and more apartment buildings, cross the street, and arrive at the stormwater retention pond. The term does it little justice: most such ponds are managed with utility in mind, ringed by turf grass mowed within an inch of its life. Not so this pond, which abuts the campus of the National Institutes of Health—the federal medical research agency—and was established under the guidance of a nature-loving landscape architect. He had its edges planted with cattails and its banks with a riot of native wildflowers. Now the pond is a pocket sanctuary.

The birdbath stream is several hundred feet away, hidden by a grove that separates the pond from a housing block. Some of the trees are quite large, first-generation descendants perhaps of forests predating the sprawl that sprouted around them. A bare upper limb on one stately oak is a red-shouldered hawk's favorite perch. Often he can be seen there, surveying the grounds, thick-bodied and broad-winged, an epitome of raptorial might.

That doesn't deter the crows. It's not uncommon to see a group chasing him through the airspace around the pond. They duel until, having outmaneuvered the hawk, the crows can veer close and pull at his wing feathers. At the tree they'll assemble on surrounding branches, cackling raucously, ostensibly to drive away this threat—although, with crows, one can't shake the suspicion that their motives are frequently beyond our ken.

Perhaps that's because it's easy to project unease onto these dark, self-possessed birds. People have done so for thousands of years, not only to crows but to many other members of their corvid genus. (You may recall that in the Bible, before sending a dove to find dry land, Noah sends a raven; unlike that obedient dove, the raven went off on his own.) Traditional stories mixed admiration with distrust, but a shift to agrarian life, which prioritized productivity and portrayed nature's residents as enemies, turned ambiguity into loathing. In the United States, people killed crows and ravens so indiscriminately that they became rare near human settlements.

Nowadays, of course, crows are everywhere. Usually that's attributed to how crow-friendly urbanized environments are; cities and suburbs indeed abound with opportunities for plucky, trash-scavenging generalists, and no doubt all that food is important. "But it's not just about scavenging," notes John Marzluff, a biologist at the University of Washington who studies crows. It's about their smarts, too.

The cognitive feats that made corvids the "feathered apes" of animal intelligence research also helped crows adapt to fast-changing landscapes. Studies documenting their abilities are too numerous to recount here—but one line of research, modeled after Aesop's fable about a thirsty crow who dropped stones into a pitcher of water, is especially instructive. In these experiments, birds are hungry rather than thirsty, and they must drop blocks into a tube of water until they can reach a floating snack. The birds quickly learn the trick.*

Scientists debated: Do the birds truly display ingenuity, figuring out what to do by applying cause-and-effect understanding, a basic grasp of physics, and imagination? Or do they rely on trial-and-error learning, association, and persistence?

The bulk of experimental evidence suggests that both approaches are involved. But regardless of the answer, it's worth reflecting on what we perceive to be at stake in the debate. Society prizes insight. It's considered more sophisticated, more *special*, than mere trial-and-error persistence. It draws upon abstract reasoning, a trait long regarded by those who denied that animals could think in meaningful ways as uniquely, even definitively human. Yet so much of what we do is trial and error: trying different things until something finally sticks. Fetishizing insight is a lingering trace of that old desire to differentiate ourselves from the rest of the animal kingdom.

The Aesop's fable studies fit a direction most research on animal minds followed until recently. Scientists focused on traits such as memory, pattern recognition, and problem-solving, which are relatively easy to measure, rather than the fuzzier domain of emotion. There *are* ways of studying emotions, to which we'll turn in the next chapter. In the meantime, let's stick to those more straightforward forms of cognition—the sort of traits that could be quantified by, say, researchers studying gray squirrels like one now chattering in a pine to see how they adapt their behaviors to changing circumstances. To investigate this question, captive squirrels can be trained to press a red rather than green circle in order to

---

* To be clear, the studies have not involved American crows, like those harassing the hawk, but several closely related species, including New Caledonian crows and rooks. Given the consistency of the results, the findings can be extrapolated.

receive a snack; then the script is flipped, rewarding a green circle-push instead. The squirrels' behavioral flexibility is measured as the time it takes them to adjust. The results are no surprise to anyone who's ever tried to keep squirrels out of a bird feeder: it doesn't take them long at all.

But other findings are more surprising. Consider the baby mallards paddling through a skein of duckweed under their mom's watchful gaze. Newly hatched chicks who've never seen *anything* before respond differently to light patterns modeled after a cat's walk than to patterns resembling a rotating cylinder; likewise they distinguish between self-propelled objects and those that are pushed. These findings suggest an evolutionarily ancient animacy detection system, primed to pick out animal movement amid swaying grasses and rippling leaves.

Many species possess this ability, not just birds—but there's something poignant about its presence in those ducklings, so fresh to this world. Other research on the cognition of newly hatched chicks found that, after following two balls hanging from a rotating boom, they subsequently preferred to follow a pair of pyramids rather than a mismatched cube and rectangle set. If one ball was red and the other green, they later chased a blue and yellow pair rather than two violet balls. The methodology might seem esoteric, but the takeaway is universal: ducklings grasped the abstract concepts of *similar* and *different*, and they could apply them generally.

Three centuries ago, philosopher John Locke wrote that "brutes," by which he meant animals, "abstract not." Nowadays the capacity for abstract thought is "often assumed to be reserved to highly intelligent organisms," wrote the researchers who tested those clever ducklings. Yet it makes perfect sense that they're so adept. Recognizing their mother and siblings is all-important. Rather than relying on a two-dimensional snapshot of mom rendered misleading the moment she spreads her wings or changes position, ducklings draw upon a mental array of characteristics and relationships that allow them to identify her. And even those animals who'd previously demonstrated this ability, the researchers noted, took some time to develop it. Seven-month-old human infants only pass the test after seeing several pairs of objects. The ducklings did it straight from the egg.

One could say they did so instinctively, although that's a word to be handled with caution. The narrow definition involves innateness: an inborn impulse or capacity, present without being learned. The scientific history of the term, though, is loaded with baggage. It's treated as mutually excluding rationality. When applied to animals, it's used dismissively. Then instinctive means *thoughtless,* the opposite of reasoned, a lesser form of intelligence than our own.

◆ ◆ ◆

UPSLOPE FROM WHERE THE DUCKLINGS PADDLE IS A stand of evening primrose. This early in the season the stalks are still short and have yet to bloom. When they do their golden blossoms will attract ruby-throated hummingbirds, and in early fall the last of their nectar will help sustain the hummingbirds on southward migrations that require a nonstop flight across the Gulf of Mexico. Regardless of whether they've made the journey before, hummingbirds bulk up before departing—but those who have previously completed the passage feed with an extra urgency, packing on nearly half their body weight in just four days. The migratory urge is instinctive, but experience has taught them the importance of carrying as much fuel as possible.

Last season's primrose stalks still stand, brown and dry, and a female red-winged blackbird is using her beak to peel off long strips for use in the nest where she'll soon lay eggs. Bird nests offer another lesson on instinctiveness and human exceptionalism. They would appear to be tools—tool use being a trait once said, before being found throughout the animal world, from chimpanzees to octopuses and ants, to make us human. Conventional wisdom says they're not.

Tools are supposed to be crafted with intent. Because individuals of some species build nests even when they've been raised in captivity, without ever having seen one, nest building is usually considered instinctive. No matter how intricate and functional nests are, their construction is regarded as the mindless unfolding of a genetic program, not a meaningful example of tool use.

Yet new studies of nest building have challenged that view. Yes, some birds build nests without prior exposure—but if they have seen one, they will pattern their own upon it, and with practice their nests become sturdier and more refined. Piece-by-piece analyses empirically confirm the impression of anyone who's studied a songbird nest up close: that the materials are not chosen and arranged randomly, but instead are selected with exactitude and functionality. The primrose strips gathered by the blackbird will become her nest's scaffold, with a mud-plastered outer cup of stiff twigs providing structural support and insulation for a soft, comfortable inner lining woven from fine grasses.

Does that mean she has an engineer's understanding of each element's structural properties? Not necessarily, at least not at first. But neither is the construction a thoughtless task. There is learning involved, and decision-making—not just about what to use in the nest, but where best to put it and how to hide it. (One of last year's nests is still semi-intact, secure in the protective brambles of a vine-tangled shrub.) Instinct and thoughtfulness blend, as they do for us. One might practice a sport until playing it is instinctive, or instinctively move to protect a loved one from a threat, but conscious thought is involved in the performance.

"We know all behaviors are a combination of those two things," explains Susan Healy, a biologist at St. Andrews University who has shown that zebra finches learn from experience. Healy points out an irony: One reason why nest building is treated as hardwired is that nests display "stereotypy," which is to say that one blackbird nest much resembles another, and a crow's nest the other crows' nests, and so on. Stereotypy is generally considered a sign of instinct—yet tools manufactured by early humans exhibit a great deal of stereotypy. "Nobody," says Healy, "thought *that* was instinct."

Birds are not the only nest makers. From a hole in the fence that separates the path from pondside vegetation sprouts an unlikely tuft of dried grass. Closer inspection reveals it to be part of a field mouse nest built mostly from tufts of honeyvine milkweed seeds. The vines themselves don't look much like their free-standing milkweed cousins, but the flat brown seeds, packed thickly into pods, are almost identical, with a shock of silky white hairs at one end carrying them almost weightlessly on a breeze. The hairs are hollow and water repellent, and there must be hundreds of tufts in the mouse nest. It looks—and likely functions—like a handful of insulation from a puffy winter jacket.

It must have taken the mouse a great many trips to finish his home: cautiously climbing down the fencepost and scurrying to the vines, pulling the hairs from a few seeds, filling his mouth and returning, and finally using the dexterous fingers of his tiny, hand-like paws to fashion the hairs into a structure. He was, perhaps, an especially industrious mouse—industriousness being among the personality traits that scientists have started to categorize.

People who spend time with animals know they have personalities, but studying them formally is a relatively recent development. Well into the 2000s, it was possible for something as self-evident as the observation that personality differences affect metabolism—that a fearful, stress-prone rabbit might tend to have a higher resting heart rate than an especially bold one—to be a novel insight. Studies of personality, defined scientifically as consistent individual differences in behavior, revolved around such general traits as boldness and shyness, risk-taking and risk aversion. The research is becoming more sophisticated. One recent review of more than 4,000 earlier studies assessed them in light of qualities such as self-discipline, tenacity, and—per our mouse—industriousness.

Our walk has taken us to the pond's far end, near the birdbath stream. Where it emerges from the drainage pipe, the builders lined the channel with concrete. Several green frogs, easy to mistake for bullfrogs but with distinctive ridges on their backs, are resting on the smooth banks, watching for flies and absorbing its sun-baked warmth.

Even George Romanes, Darwin's protégé and someone so credulous about anecdotes of animal intelligence that he eventually gave their study a bad reputation, wrote that "on the intelligence of frogs and toads very little has to be said." That attitude lingers today, with frogs and toads—and indeed amphibians and reptiles in general—receiving little attention compared to birds and mammals. Yet what attention they have received reveals creatures of considerable cognitive sophistication.

Research shows that the tadpoles of common wood frogs, another frog species found in the pond, extrapolate from past experiences with predators to estimate future risk. They also respond to the threat of predation less warily than do green frogs, perhaps because they evolved to lay eggs in ephemeral ponds; tadpoles must grow up fast, before the waters dry out, and across evolutionary time it's been worth the risk to keep foraging even when danger is near. Tellingly named edible frogs quickly learn to discern the odors of different predators. Italian treefrogs can count up to two and also distinguish two from four, drawing on a basic numerical sense that's believed to have evolved hundreds of millions of years ago. It likely helps them choose between habitats with differing amounts of vegetation. Other frogs whose courtship rituals involve a back and forth between males competing vocally for attention can count as high as eight, helping them track the croaks they must make to top competition.

These studies don't capture the entirety of frog cognition. Rather, they suggest what might be learned when people are attentive to them. One study has even focused upon the personality of American bullfrogs; it was limited in scope, but suggested they vary most in their tendency to explore. Perhaps some of those frogs on the drainage channel's concrete banks will be content to stay near home. Others may feel a yearning for more.

◆ ◆ ◆

FOLLOW A SIDE TRAIL THROUGH A DENSE STAND OF honeysuckle and we end up inside the grove, where the canopy casts a green tint and several boulders make good seats. The ground is covered by violets on their last blooms and English ivy. It's an invasive plant, but the fat-rich berries are a winter staple for local animals and the year-round foliage a good source of cover. From beneath a leaf comes a rustle and a momentary glimpse of whiskers, nose, and eye, wary but curious.

A brown rat. It'd be nice if he came from beside that restaurant up the street, his family having tunneled their way around that concrete entombment as surely as they avoided years of poison and traps. Unlikely, but at least possible. *Rattus norvegicus*—migrants from southern China who, despite never-ending campaigns of extermination, are now as widespread as we are—are nothing if not resilient.

To those few people who keep rats as pets, they're also sweet, smart beings, not so different from dogs except even more predisposed to gnaw on things. To scientists, they're a model organism for studying the basics of brain and biology; we probably know more about rats than any creature other than humans. Depictions of them in popular culture are often sympathetic, like the movie *Ratatouille* and its five-star rodent chef or viral sensations like Pizza Rat, onto whose struggles to carry a slice up the steps of a New York City subway station the human condition could be projected.

All that has not yet translated into much consideration for rats.

In cities, poison bait boxes are so ubiquitous as to be invisible—although, as we'll see later, that mentality is starting to change, with the reformation of so-called pest control. For now, consider one particular insight: that rats are capable of yet another ostensibly unique human capacity, what scientists call mental time travel. They not only recall the past—something many animals do—but can imagine the future, an ability assessed by making recordings of rats' brains as they navigated a maze and then later at rest, when their brain patterns showed them mapping out alternative routes.

Mental time travel is part of the autobiographical sense of self that's so critical to our own self-awareness—the sense of oneself as an entity, distinct from other individuals and from one's own environment. It's so intertwined with our own experiences that it's practically impossible to imagine *not* being self-aware. In everyday life, we implicitly think of other animals as having a sense of self, too. How could a blackbird finish building a nest except with some understanding that it's *for her*? Or a squirrel cache an acorn for future eating? Lacking a sense of self, they'd be like windup toys going through the motions of life.

Yet self-awareness is a thorny topic in scientific circles.* Only a few animals other than *Homo sapiens* are formally recognized as having it, among them chimpanzees, orangutans, bottlenose dolphins, Asian elephants, and—depending on whom you ask—magpies and manta rays and cleaner wrasse fish. They've all passed what's known as the mirror self-recognition test, developed in the 1970s and since then treated as a gold-standard measure of self-awareness. To pass the test, an animal must use a mirror to investigate a mark on their body that can't otherwise be viewed directly, such as paint dabbed on the back of their head while they slept. It can be inferred from their curiosity that the

---

* By self-awareness I mean an awareness of oneself as an entity, rather than the basic it-feels-like-something-to-be-that-entity quality of sentience.

reflected image is incongruent with their own mental self-image. They're self-aware.

Yet many scientists now argue that too much emphasis is placed on the mirror test. Impressive as passing is, failure doesn't necessarily signal an absence of self. (Indeed, human children in Western societies tend not to pass the test until they're about 18 months old, and children in Asian and African cultures at 5 years of age. Nobody would suggest they lack self-awareness.) It could be that sight isn't so relevant to other species' self-ness, and new variations of the test use other sensory modalities, such as odor. One suggestive experiment found that dogs are especially interested in their own scent when it's been mixed with another, unexpected smell; a cousin to this experiment, performed with garter snakes like those sometimes glimpsed vanishing into the high grass, found that they distinguished their own scent from that of their siblings. Their self-image may not be an *image* at all.

The mirror test is also criticized for promoting a binary, all-or-nothing conception of self-awareness. As animal behavior researcher Frans de Waal wrote of a controversial experiment in which cleaner wrasse seemed to recognize themselves—arguably their behavior was ambiguous, but more than that, how could such a tiny fish possess such a sophisticated capacity?—self-awareness is viewed through something like "a 'Big Bang' theory, according to which this trait appeared out of the blue in just a handful of species.'" Instead, suggests de Waal, we should think of self-awareness developing "like an onion, building layer upon layer, rather than appearing all at once."

When one teases apart these cognitive layers, they're not so uncommon. Mental time travel is one of them. Another is metacognition, or the ability to reflect upon one's knowledge: *I know. I don't know.* Rats are also metacognitive, as demonstrated in experiments that let them choose between getting a small, guaranteed snack now or taking a memory test in which correct answers earn big snacks and wrong answers nothing at all. When the passage of time has blurred their memories, they opt for the guaranteed reward.

Some might argue, though, that rats are uncommonly intelligent, and therefore not representative. Yet consider another component of self-awareness, episodic memory: the *what, where,* and *when* qualities that give shape to undifferentiated recollection. Episodic-like memory has been found in zebrafish, a tiny species used as a model organism for investigating the foundations of cognition. Minnows shoaling in the stormwater pond's shallows almost certainly share this type of memory, with the topographies of their daily lives replacing the experimental setups—familiar and unfamiliar objects, familiar and unfamiliar settings—used to illuminate the memories of their aquarium-bound cousins. Researchers have also demonstrated this type of memory in hummingbirds, mice, and cuttlefish, the latter of whom last shared a common ancestor with vertebrates more than 500 million years ago. That such evolutionarily disparate creatures possess these memories suggests how common they are.

Instead of some self-awareness being rare in the animal kingdom, then, forms of it are widespread—not so complicated as our own, perhaps, but still meaningful. And some scientists argue that the essence of self-awareness is truly simple indeed. They distinguish between the sort of contemplative self-awareness measured by the mirror test and the baseline self-awareness produced when a brain integrates perceptions from different sensory organs with an internal representation of their environment. Distinguishing oneself from everything else is the basic ability required to process new information provided with each step, wriggle, or wing flap.

By that light, self-awareness isn't just widespread; it's ubiquitous. Forms of it have existed ever since some tiny worm-like creature, navigating warm Ediacaran seas before even plants had evolved, could perceive both light and contact, compare their present state to a mental representation of the immediate past, and act.

Light was here, now it's there, so move. My body did this, now it does that. In that simplicity may reside the origins of self.

It's why an earthworm responds differently to the touch of soil disturbed by your spade than to the touch of soil pushed by its body. It's evident in an experiment in which recently fed rat snakes, their bodies dramatically swollen, enter their enclosures via large holes rather than the smaller holes used prior to feeding. It's there in the movements of jewel-eyed jumping spiders patrolling grasses along the pond's path, tracking out-of-sight prey, and pouncing to where they'll be in a moment's time. At some level they're aware of themselves. "To have a thought, experience, or emotion," writes philosopher Mark Rowlands, "is to be aware that this state is mine." Every single creature you encounter is a *someone*.

Set back from the path is a wooden bluebird box now occupied by a pair of black-capped chickadees. Come autumn they'll collect thousands of seeds, their larder for the coming winter, and hide them under pieces of flaking bark on the branches of small trees. The extent to which they're consciously aware of winter is unknown; studies show they can plan at least a half-hour ahead, but months may fall outside their cognitive limits. Instead chickadees may simply have some inkling of future, and that is enough. Right now, they're concentrating on feeding their nestlings, who erupt in a chorus of eager trills each time mom or dad arrives with a mouthful of bugs.

Their self-awareness is certainly not identical to our own. Self is shaped by those many layers of cognition and sensory perception. But there is a commonality. If we could enter the mind of one of those chickadees, a frog on the bank, the ducks, the rats, and even the minnows, we would be able to relate.

2

Landscapes of
the Heart

AFTER THE MORNING RUSH THE POND FALLS QUIET until early evening, when a new bustle of activity greets the waning light. Cottontail rabbits emerge from long grass to nibble in the walking path. They are subtly beautiful creatures, with brown fur that appears uniform from a distance but in proximity is a filigree of hues. Their big black eyes see in every direction; their tall ears are parabolic microphones on swivels. These are the adaptations of a supremely wary creature. That they're so unbothered by pedestrians, hopping just a few feet off the path when someone approaches, feels like a gift of trust. Perhaps the taste of fresh clover and dandelion influences their reluctance, too. It's always hard to get up from a good meal.

A short way up from the birdbath stream, behind the wrought iron fence that surrounds the NIH campus, is a large black willow tree—the largest of its kind in the county, standing 60 feet tall with a multi-trunked canopy that's nearly as wide. At the base of a limb overhanging the path is a hole where a mother squirrel made her nest. With the day's heat past, her kids come out to play. They chase each other around and through the branches, leaping between them, hanging upside down, and every so often pausing for a nuzzle before starting up again. They look as if they're having a blast.

It's strange to think that, until recently, ascribing pleasure to the squirrels' antics or gustatory delight to a dandelion-savoring cottontail would have been a controversial statement among many scientists. They would have met the claim with skepticism: Could we ever truly know if a squirrel is having fun or if a rabbit is enjoying a meal? Though they give the appearance of it, are animals even capable of those feelings as we understand them?

Even as science opened its mind to the possibilities of animal intelligence, the subject of animal emotion was kept at arm's length. Scientists did so in part because of the challenges of studying emotions in creatures who couldn't talk about them and in part because of how people think about emotions in the first place. For centuries scholars drew a sharp contrast between intelligence and emotion, with the former held in higher regard. Modern conversational habits still reflect this legacy. Using your head is the opposite of following your heart, as it's been since Plato located rationality and passion in different parts of the body; brainteasers don't involve feelings. The title of genius is rarely bestowed upon people of exceptional empathy. There's a gendered tinge to this, too: rationality is a historically masculine trait, whereas emotions are unreliable and feminine, and scientific institutions have an unfortunate history of misogyny.

Perhaps animal emotions made people uncomfortable in other ways, too. The theory of evolution, with *Homo sapiens* merely one branch among many on life's tree, was disconcerting enough to believers in human exceptionalism. If scientists might reluctantly admit that a mother squirrel could possess insight or memory, the notion of her sharing our *feelings*—that the affectionate exhaustion aroused by her rambunctious kids might resemble a human mom's—was just too personal.

Some scientists and philosophers argued that experiencing emotions in any meaningful way required language. Without a word for love, its raw physiological sensations could not be transmuted into conscious feelings. This assertion was speculation, though, and assumed a far starker divide between human and animal communication than actually exists. Others didn't go so far as to call language a requirement, but did argue that, in its absence, we simply could not determine in a rigorously scientific way what animals felt.

This view was understandable. Emotions are subjective; unlike traits such as learning or memory that can be measured with stopwatch certainty as an animal finishes a maze or solves a puzzle, emotions can't easily be quantified. In studies of humans, scientists resolve this ambiguity by asking people how they feel. That's not possible with animals, and it's easy to make mistakes. Darwin wrote in *The Expression of Emotion in Man and Animals* about a Sulawesi macaque monkey expressing affection with a tooth-baring grin. That's now understood as a grimace of fear or submission.

Nevertheless, the reluctance to acknowledge emotion in animals will probably seem strange to anybody who has lived with one. Outside academia a great many people never had any doubts at all. Hence the rich body of folk stories, entertainments, and personal accounts—from *Lassie* to *Gentle Ben* to *March of the Penguins*—that helped shape modern Western culture. Yet these observations were not formal knowledge, and in serious settings scientists dismissed that kind of thinking as anthropomorphic sentimentality.

Several currents would combine to erode that reluctance. A deeper scientific appreciation of human minds, as well as the neurobiology that gives rise to our own emotions, has shaped research on those of animals, which in turn yielded more insights into us. With this came a new sense of cognition and emotion—thought and feeling—as intertwined, even inextricable. Take emotion out of the mental equation and ostensibly abstract processes such as reasoning and problem-solving are crippled. Some scientists even suggest that the systems underlying emotion evolved with the ear-

liest vertebrates. Feelings helped creatures react to changing circumstances and pursue useful behaviors. The rabbits by the path
have a felt incentive to eat dandelions.

Combine those insights with clever new study designs and the
richness of animal emotions becomes unavoidable. Yet even as scientific journals now overflow with such studies, their lessons have
percolated unevenly into our awareness of nature. We read about
how voles are used to study the neurobiology of love, how salmon
become frustrated, how even crawfish experience anxiety—but of
course this doesn't happen only in experimental settings. Earth's
lands and waters and skies teem not only with intelligence, but
with feeling.

◆　◆　◆

WATCH THOSE COTTONTAILS EATING GRASS, THEIR
sensitive noses slowly twitching: they're not merely acquiring nutrients, but enjoying a meal. The pleasure of eating is evolution's
way of encouraging us to eat well—and if there isn't yet a study on
cottontail gustatory affect, of what it's like to eat with three times
as many taste buds as humans have, we might look to a suggestive
study on rats who braved a 50-foot dash through freezing temperatures in order to eat shortcake and soda rather than their usual lab
chow. Taste is one of the eternal satisfactions, and there's something vicariously satisfying about watching rabbits munch.

As for those acrobatic young squirrels in the black willow tree,
there's plenty of scientific research on the importance of play
among animals. It helps them—and us—learn about the world:
physical laws and properties, the features of their environments,
how their own bodies work, and rules of social etiquette. Like all
the best teachers, evolution encourages play by making it fun.

One of the squirrel pups likes to hang upside down; he acts like
a kid on a jungle gym. A little flush of dopamine should reward that
squirrel's feats, leaving him wanting more, and so the pups leap

from limb to limb, chasing each other round the trunk until they're too tired to continue. Such play—which scientists formally define as voluntary, repeated behaviors that serve no obvious function, are dissimilar from "serious" behaviors, and occur in nonthreatening situations—has been documented not only among the usual suspects (*horsing* around) but throughout the animal kingdom. Vietnamese mossy frog tadpoles catch rides on bubbles; elephant fish balance twigs on their snouts; Komodo dragons lumbering after a ball resemble, when viewed on sped-up video, frolicking dogs.

While the squirrels play in the lowering sun's long light, a pair of starlings pay a late-day visit to the birdbath stream. They dip their heads and beat their wings, sending up a spray that soaks their plumage and deepens its breeding-season iridescence. Starlings don't receive nearly enough credit for being beautiful. At this time of year, the nominally black feathers of breeding adults ripple with purple and emerald, while the tips of each nonbreeding adult's feathers are dipped in gold.

Several years ago, a British ethologist named Melissa Bateson conducted a series of experiments designed to learn about how bathing makes birds feel. She adapted so-called cognitive bias tests originally developed to study humans, particularly very young children and other people who couldn't easily articulate their feelings. These rely on the way moods shape responses to uncertainty: when we feel good we're optimistic, seeing the proverbial glass as half full, and are more likely to take chances. When we feel bad the opposite is true. The same holds for animals, too.

Bateson made recordings of starlings crying out in alarm and played them back to the birds in her lab. Those who bathed earlier that day looked up, scanned their surroundings, and quickly returned to their meals. They were in a positive frame of mind. Those who had not been allowed to wash, however, were slow to resume eating and remained vigilant throughout. They were anxious; the possibility of being attacked seemed to loom large. The starlings in the stream, at least, should be plenty relaxed.

Cognitive bias tests have been administered to many species both wild and domestic. They explore, among other matters, the psychological effects of close confinement on pigs, whether sniffing makes dogs happy, and how ravens become upset when seeing another raven in a bad mood. Bateson is among the researchers who even refined these tests for bees. Honeybee drones are less likely to investigate an ambiguously bittersweet scent after their hive is shaken, evidently because being attacked puts them in a mood—a mood being, to be precise, a lingering emotional state rather than a fleeting feeling. Other researchers did similar work with bumblebees, first training them to associate artificial blue flowers with sweet nectar and fake green flowers with tasteless water, then presenting them with blue-green flowers of uncertain reward. Rather than upsetting them, though, the researchers gave some of their bees a sugary treat. Those bees flew quickly to the strange flowers. They were in a good mood.

Add enough positive experiences and good moods together and, in animals as in *Homo sapiens*, we get a general state of happiness—which, even for people predisposed to recognize animal emotions, seems a stretch when applied to bees. Yet those tests of optimism and pessimism are just one of several lines of findings that point in the same direction. When researchers dose bees with chemicals that block the action of serotonin, octopamine, and dopamine—neurotransmitters that regulate emotions in humans—the bees no longer act emotional. And when the brains of bees and other insects are examined closely, they prove to have brain structures and chemistries that resemble our own so closely that researchers now study invertebrates in order to develop a better understanding of human emotional disorders. Other research has described how fruit flies enjoy ejaculation and, when subject to repeated, uncontrollable stresses, become depressed. Even the invertebrate world is full of feelings.

At the end of the pond opposite the birdbath stream, the slopes are planted with swaths of beautifully named foxglove beardtongue. The "foxglove" is perhaps an old English derivation of *folk*, meaning faeries, their hands small enough to fit inside the blooms; the "beardtongue" comes from the fine hairs that line each blossom's lower petal. In spring they swarm with carpenter bees and bumblebees, each crawling inside to sip nectar and receive a lick of pollen from those hairy tongues. They come and go in such numbers that their buzz is audible from a blanket laid on the grass beside them. That buzz is the sound of the pleasures suffusing this place.

The obverse of pleasure is pain, that union of sensation and emotion: not only the pressure of a stubbed toe, but the emotion that makes it an unpleasant experience. Until fairly recently, it was argued that large-brained creatures feel pain more intensely than small-brained creatures and that only humans and perhaps other mammals could even feel it in any meaningful way. Those arguments no longer hold much scientific sway, but that they ever did is telling. It perhaps says less about what was scientifically known than how difficult it can be to empathize with creatures physically different from ourselves.

Particularly instructive is an argument that continues even today around fish, whom I grew up catching with the understanding that, however much they might thrash when impaled on a metal hook, they didn't actually feel pain. It was supposedly something called nociception—a purely perceptive response without any emotional dimension. This idea was based largely on the belief that fish lack brain structures that regulate pain in humans.

That view too has fallen out of favor. Too many studies show that, exposed to something we find painful, fish behave as we would. They change their behaviors in order to avoid repeating the experience, suggesting a psychological dimension that encourages them to learn and adapt. As with bees, giving fish pain-blocking drugs short-circuits those reactions. Their brains, and for that matter all vertebrates' brains, also have analogues of those human pain-regulating structures—not exactly like our own, but close enough. Evolution found ways to produce

similar experiences using a variety of forms. Pain, after all, is extraordinarily useful.

This doesn't mean fish pain—or, for that matter, pleasure—is precisely like our own. There are some obvious differences. The pain-sensing neurons of trout, for example, don't respond to frigid temperatures beneath wintertime ice, yet their easily damaged skin is actually more sensitive than our own. Our own emotions should also not be seen as defining the spectrum of possibility. Becca Franks, a psychologist at New York University who specializes in fish cognition and welfare, offers the example of an Atlantic salmon returning from the ocean to spawn. What might the currents of their home stream *feel* like? Not just against their bodies but, so to speak, in their hearts.

It's not wrong to look at frogs basking in the sun and imagine them content—yet is that all? Many amphibians can perceive faint electric fields, and each winter their bodies slow to a near stop for months at a time; these physiological feats are beyond our ken and make one wonder about emotional analogues. What feelings might exist for which we don't even have a reference?

◆　◆　◆

AS EVENING FALLS A LONE MOCKINGBIRD SINGS FROM a fencepost above the beardtongue. A male, it seems, from the sheer variety and volume of his repertoire. Females also sing—a fact that, not only in mockingbirds but in a great many bird species, is underappreciated—but somewhat less boisterously, and not so much in spring. His feathers are dull gray and white but his eyes are a fierce yellow that befits the passion of his performance, which isn't so much a song but rather a concert, like a jazz musician riffing a medley of popular tunes or a DJ mixing records. It lasts for eight or ten minutes before he flies to another fencepost and begins anew.

In the medley are snatches of red-winged blackbird trills and red-tailed hawk screeches—or maybe it's blue jays imitating hawks—and catbird mews and crow caws, calls learned from dozens of species and surfacing faster than an untrained ear can follow. "Whate'er birds did or dreamed, this bird could say," wrote the nineteenth-century poet Sidney Lanier, one of many to celebrate a species dubbed by Walt Whitman "the American mimic." Mockingbirds also make their own, original melodies, an ability that is obscured by this penchant for mimicry. In New Orleans, disagreement exists as to whether they imitate the coda played by musicians to commence street parades or if it was the musicians who borrowed the mockingbirds' song.

But why do they sing in the first place? Among the usual explanations, one is to stake a territorial claim. I like to think of this as beating their bounds, a millennia-old tradition from England and Wales in which villagers walked the edges of their lands each spring, singing all the while. The mockingbird on his fencepost may also be trying to attract or keep a mate; their unions last for years, sometimes for life, and males will serenade their partners throughout the year. Yet the language of territory and mate attraction can seem mechanical, even dismissive. They treat singing as rote. These explanations don't account for what it feels like to sing. Might this mockingbird experience some of the joy that a human performer would?

The short answer: He does, but it's a bit complicated. Research on zebra finches—a species sometimes used to draw general conclusions about how songbird brains work—suggests that when the audience is a mate or prospective partner, songs are motivated by pleasure seeking; after sex, males stop singing, ostensibly because they've been gratified. Scientists test this experimentally by dosing male zebra finches with chemicals that amplify the opiates circulating naturally in their blood, mimicking post-coital bliss. Afterward their female-directed songs are quiet and halfhearted. Already sated, they don't need more satisfaction. The rewards of territorial singing also seem to reside in their consequences rather than the performance itself.

Yet not all singing is about territory or mating. Juvenile birds sing long before these concerns ever enter their feathered heads, and as adults both males and females often sing in the absence of potential mates or competitors. These tunes are ignored by other birds. They seem to be performed for the singer's own benefit, for reasons suggested by studies of the role of dopamine in song learning. As birds sing, they make a mental comparison between an idealized version of the song that exists in their heads and the notes coming out of their beaks. A match yields a neurotransmitter flush. Simply hitting the right note is its own reward.

And what of the experience of listening? Particularly when it comes to courtship, scientific descriptions come up short; females might tend to favor those males whose vocal qualities correlate with reproductive success, but those patterns play out across evolutionary time. They don't capture what it feels like to hear a song. Here some insight comes from a study of white-throated sparrows, infrequent visitors to the pond, who winter in the southern United States but breed in the conifer forests of Canada and New England. Their high, clear whistles suggest crisp mornings and brooks bubbling over tree roots. When female white-throated sparrows listen, the same neurological pathways become active that in humans respond to music.

◆　◆　◆

THOUGH DARWIN WAS NOT THE BEST INTERPRETER OF monkey expressions, he was a visionary on the subject of animal emotion. He saw no reason why feelings should be unique to *Homo sapiens*; instead ours could be traced to basic capacities—joy, despair, determination, anger, disgust, and surprise—that he believed to be widespread among animals. So deeply rooted were emotions in evolutionary history, surmised Darwin, that the very act of vocal communication was their legacy, bequeathed when spasms of excitement caused some primeval air breathers to contract their windpipes, producing inchoate sounds that natural selection would tune into frog ribbits, alligator roars, and human language.

Natural selection would also shape the ability to perceive these sounds. If emotional expression indeed derives from a common root, then humans should be able to interpret sounds of arousal in very different creatures—a proposition tested several years ago when researchers at Germany's Ruhr University Bochum had people listen to recordings of nine species, among them representatives of mammals, birds, reptiles, and amphibians, and guess whether the animals who made them were excited. Regardless of the species, people got it right. One species was the humble black-capped chickadee, our future-planning friends from the previous chapter; people actually recognized their arousal more readily than that of pigs or Barbary macaques, who are far more closely related to us.

Other researchers analyzed sounds of distress made by infants from every kind of vertebrate except fish and found common acoustic signatures. The leader of that work, a biologist named Susan Lingle, has also conducted experiments in which recordings of distressed baby mammals, including marmots, seals, cats, bats, and *Homo sapiens*, are played for mother deer. The deer moms responded to all of them, even to us. A baby's distress is universal.

Findings like these underscore not only Darwin's insightfulness, but also a connection between emotion and social life so powerful that its purest expressions span evolutionary gaps of hundreds of millions years. If emotion started as a way to help creatures adapt to changing circumstances, it soon became central to navigating social interaction; in every vertebrate lineage, the neurobiological networks that regulate emotion and govern social behaviors are intertwined.

Emotion is profoundly social, and wonderful ambassadors for this principle are none other than our neighbors from the restaurant sidewalk hole and the pocket forest ivy patch, *Rattus norvegicus*. In the last decade scientists have performed experiments showing how deeply rats feel for one another. They respond to the distress of trapped cage mates with such urgency that they'll forego chocolate—a delicacy for rats as well as for us—in order to save them, and they are especially generous when sharing food with rats who are anxious.

These experiments both cast a new light on a much maligned creature and illuminate the nature of empathy itself. Rats don't possess the highly abstracted ability to imagine oneself in another person's place, even when having learned of them through a secondhand account; that, perhaps, is unique to humans. But there are simpler forms of empathy, the sort displayed by rats, as when someone is crying and you can't help but feel upset yourself.

Such "emotional contagion" is thought to have evolved to foster sociality, particularly in those species whose mothers care for their young and must be attentive to their moods and needs. Rats do this, mothers nursing and grooming their pups in dens hidden from our sight, the pups making sounds of joy too high pitched for us to hear. And, of course, parenting is not restricted to mammals but is found throughout the animal kingdom, even in some species of insects. Burying beetles who cache the dead bodies of small rodents and birds regurgitate food for their begging offspring; during this time their brains undergo changes in chemistry that are associated with parental care. Before long that mockingbird and his mate will be feeding a fledgling grown nearly as large as they are, so that when he opens his mouth to beg it looks almost absurd.

Once the neurobiology of empathy has evolved to encourage parental care it can be turned to other relationships, especially those between mates. There's nothing better than shared feelings to unite individuals through life's tribulations. A few blocks from the pond, along a walking path built on an abandoned railroad line, is a colony of pine voles. These are chunky brown relatives of house mice, whose prairie vole cousins have achieved some scientific fame in studies of oxytocin, a hormone central to the mental processes underlying reconciliation and affection between partners. Oxytocin evolved about 450 million years ago and plays a part in the behaviors of every vertebrate class, but prairie voles have provided a unique window into its influence on choosing mates, strengthening relationships, and expressing care. Sometimes oxytocin is even called the "love hormone"—a term that's much too simplistic, but does convey its importance.

Yet for all that research on prairie voles and other animals has taught us about how oxytocin works, *love* is a term mostly denied to them. The language of romantic relations among animals is decidedly sterile: whereas we have partners, lovers, and spouses, they have mates. Pair bonds and fitness benefits replace passion, romance, and good old-fashioned lust. Their choices are too often described as though animals choose partners by calculating the spread of their genes across time.

It doesn't help that animal relationships are mostly hidden to us. Farmed animals are mostly kept behind closed doors and don't live in natural arrangements. Pets come from shelters or breeders. Wild animals, with a few exceptions, are glimpsed rather than followed intimately across days and seasons. The most detailed accounts come from viral videos or the handful of people whose lives allow them to observe animals up close.

Rita McMahon, the founder of City Wildlife, a hospital for injured wild birds in New York City, once told me about caring for a pigeon with a broken leg. While the bird recuperated on a cushion in her window, the bird's mate would stand on the other side, keeping her company until finally she was released and the couple could reunite. Kevin McGowan, a biologist at Cornell University who has tracked more than 3,000 crows over the last three decades, describes watching a crow named AP choose between two females; the one he rejected became very successful with another male, raising brood after brood, while those of AP and his chosen partner failed every year.

Despite that they were a close, attentive couple, recounts McGowan, bad with fledglings but good with one another, staying at one another's side for eight full years. They preened each other regularly—a behavior that "has been hypothesized to play a role in strengthening and maintaining pair bonds," wrote researchers in a review of this behavior in birds, much as countless caring touches deepen the affections that sustain our own relationships. When AP's mate died, he lost his territory and spent the twilight of his life scavenging at a local compost facility. McGowan wonders whether AP, his heart broken, simply thought, "She died. Why do I even want to fight for this territory anymore?"*

Canada geese are another devoted species, taking mates when they are several years old and often remaining with them for life, which may span a decade or two. They are frequent visitors to the stormwater pond. Nesting is discouraged by dense cattails, but in late winter, at the cusp of the transition to spring, dozens will gather on the water. They turn somersaults to bathe and some perform elaborate courtship rituals, with couples repeatedly dipping

---

* Even as corvids have become darlings of animal intelligence research and popular science, the emotional aspects of crow life still get short shrift. When a study on so-called crow funerals—where dozens or hundreds of the birds gather around a dead crow—suggested that they learn of potential threats by inspecting the body, the finding was framed in the popular press as debunking the idea that crows mourn. The research did not show this, though.

"Danger learning is certainly one part of this behavior, but not necessarily the whole," says Kaeli Swift, the study's lead author. Crows in the experiment were presented with unknown corpses; they may have reacted differently to deceased family, friends, or partners. That sort of experiment, of course, would likely be quite unethical, and the episode speaks to the difficulties that still exist in studying emotion. "Any scientist who works with crows is thinking about this a lot," Swift says. "We just don't have great ways to ask these questions."

their heads while facing one another so that their necks, seen in profile, trace the shape of a heart.

Such rituals, which academics describe in terms of "inter-individual coordination," are not the sole province of geese seeking mates; they also help to maintain relationships. While there's not much research on the emotional lives of Canada goose couples, there is a quite a bit on greylag geese, a closely related and similarly monogamous species. Among greylags, individuals in long-term relationships are less stressed by fights with other geese; they soothe one another and become upset when separated. Konrad Lorenz, a Nobel Laureate ethologist who studied their relations, had no qualms about equating their grief to our own. "Geese possess a veritably human capacity for grief," he wrote, "and I will not accept that it is inadmissible anthropomorphism to say so." Lorenz observed geese whose mates had died become listless, lethargic, and unwilling to eat—the telltale signs of human depression, which could last for months.

Sometimes widowed geese take new partners. Sometimes they do not. A Canada goose who returned daily for three months to the spot where a car struck her partner outside a shopping mall in Atlanta, between a Papa John's Pizza and a First Payday Loan, became a local celebrity before being adopted by local animal rescuers. That story was more remarkable for its setting than the fact it happened. Internet message boards abound with stories of bereaved goose mates unable or unwilling to love anew. Once again, the scientific explanation of those broken pair bonds is that they ensured closer cooperation in raising goslings, thus propagating their genes—but none of that explains what they feel.

Of course, those examples involve relatively long-lived, monogamous species, a life history that is both fertile evolutionary soil for intense feelings—they strengthen the partnership—and relatively uncommon in the animal kingdom, albeit less so among birds. It would be inappropriate to look at every mating arrangement this way, although similarly inappropriate to think nonmonogamous animals lack any feeling for their partners. But those quibbles aside, the geese in the pond can be viewed another way: as symbols of the feelings that course through all creatures.

And what draws any two geese together in the first place? How did those couples performing their aquatic dances in the storm-water pond choose one another? The basic explanations, for geese and for other animals, involve aesthetic properties linked to reproductive success: big bodies, healthy feathers, feats of physi-cal coordination. Yet as with taste and nutrition, benefits are not rationally computed, but shape a subjective response. What is good for a rabbit to eat also tastes good to the rabbit—and some scientists now suggest that a similar dynamic holds for aesthetics. Qualities important to survival feel good to gaze upon, an emo-tional phenomenon at the heart of the property we call beauty. A goose looking at a mate, a squirrel at an old hollow tree, a bumble-bee at a beardtongue blossom: they may well find them pleasing to the eye and live in a world imbued with satisfaction.

3

No Animal Is
an Island

THE POND'S RESIDENT SONG SPARROWS CAN OFTEN be seen swaying on stems of tall grass, gathering seeds. They are subtly handsome birds, drab and difficult to identify from a distance—birding enthusiasts sometimes classify them among the LBFs or LBBs—but closer inspection reveals a streaked plumage of chestnut and cream.

The first time I spotted one, I reflexively identified him by type: an exemplar of a song sparrow. Later I'd appreciate the sparrows as individuals and neighbors, but vast biographical gaps remained. I caught only glimpses of their lives. Their social world stayed mostly hidden. That's often the case with wild animals, even those who live beside us. Identifying each individual creature is challenge enough, much less tracking them across months and years, painstakingly recording their interactions.

For some song sparrows, though, biologists at the University of Washington did just that. They put colored plastic bands on the legs of 80 juvenile sparrows living in Seattle's Discovery Park, making each identifiable by sight, then watched them through binoculars over the course of a year. It's one of my all-time favorite studies. There's something wonderful about how the social network diagrams and movement records revealed a hidden dimension of everyday life.

Having fledged and left their parents nests, most young males partnered with other young males, spending much of their summer and autumn days exploring the park together. Some ranged a quarter-mile or more within several feet of each other—quite the journey for creatures who can fit in your palm and who opened newborn eyes but a few weeks earlier. Come spring they found mates and then established adjacent territories, weaving grass nests near the same sparrows they'd grown up with.

"Many of our subjects formed close and long-lasting associations with specific individuals," the researchers later reported. They described their findings as "surprising." The sparrows' associations didn't seem linked to mating, habitat, or food. Instead, the researchers surmised, these bonds could help young birds learn songs and defend their homes. The word "friendship" did not appear in their speculations.

The omission is unremarkable but worth examining. Even while people, including scientists in their domestic lives, readily describe as friends two dogs who seek each other out at the park, and interspecies animal friendships are viral video standbys, science is cagey about the term.

That may be because enjoying someone's company is central to our concept of friendship, but studies like the one of Discovery Park's song sparrows measure only behavior, not subjective experience. Speculating about the birds' feelings goes well beyond the data. It's a fair misgiving, although it's equally fair to note that many other animals possess the capacities of memory and affection that underlie human friendship—so the potential is certainly there. Scientists also tend to explain behaviors in terms of reproductive fitness-enhancing evolutionary strategies, not feelings—

but, as we saw in the last chapter, feelings motivate the behavior. Our own capacity for friendship conferred benefits across evolutionary time, but that's not why we like hanging out.

Reticence about animal friendships can feel like dogma or unexamined bias rather than intellectual rigor. Consider the experience of anthropologist Barbara Smuts, who after finishing her doctoral thesis on baboon relationships found herself in a confrontation with a high-profile academic publisher who refused to use her chosen title of *Sex and Friendship in Baboons*. Never mind that Smuts spent two years closely observing a baboon troop, traveling with them daily from dawn until dusk, and considered friendship both the most plausible explanation for their behaviors and essential to understanding their reproductive strategies. The publishers insisted on putting friendship in quotation marks. Smuts refused.

"Not to call them friends seemed to be taking something away from who they really are," explains Smuts, who opted for a smaller publisher who let her keep the title. That was in 1985. Science has come a long way since, Smuts says, and her views are not considered quite so radical now. Some scientists are now comfortable with the language of friendship. Even so, the term usually attaches to charismatic, high-profile species such as elephants, dolphins, and chimpanzees, whose widely celebrated intelligence makes it uncontroversial to speak of their human-like properties.

Meanwhile sophisticated tracking technologies have illuminated relationships throughout the animal kingdom. Female barnacle geese, for example, prefer the company of certain other females when they're young; these associations dwindle as they find mates, but rekindle after breeding season. They're native to the Arctic, but share a basic life history—long-lived, monogamous, highly social—with many goose species elsewhere, including the stormwater pond's Canada geese. The lady geese performing courtship displays in springtime and later leading trains of fuzzy, awkward goslings will soon resume their grown-up friendships.

Unlike the monogamous geese, starlings who gather in early

evening on the oak where red-shouldered hawks like to perch tend to be polygynous, meaning that two or more females mate with one male. Outside of breeding season, female starlings form close relationships with other females; they spend most of their time together and share most of their songs. During breeding season these girlfriends share a mate and live in neighboring nests, too. Should their male try to bring a new companion into the picture, they'll give her the cold shoulder.

Such findings add a new resonance to the welcome return of warblers in springtime, arriving en masse as trees leaf out, brightening mornings with their variety: are their continent-spanning, many-thousand-mile journeys not only feats of navigation and physiology, but undertaken with friends? Few studies have explored that question, but one spectacular account of migration involves European bee-eaters, a rainbow-hued songbird who lives in groups that remain together almost constantly, sharing summer breeding and winter foraging areas and each year flying some 8,900 miles round trip. Sometimes they become separated on the journey and can even end up thousands of miles apart, but most eventually find each other again.

"Being together is intentional, suggesting strong social bonds," says Kiran Dhanjal-Adams, an ecologist who tracked the bee-eaters' sociality. "That they reunite, despite separations, not with other individuals but with their flight buddies, also suggests strong social bonds." As with the song sparrows in Lincoln Park, Dhanjal-Adams and colleagues didn't study the bee-eaters' feelings, but one can imagine they felt joy upon reuniting.

Research also opens possibilities for species not typically considered social, at least not in meaningful ways. The path on the abandoned railroad line that passes by the amorous vole colony leads to Rock Creek Park, where painted and red-bellied turtles can be found basking on half-submerged logs. Friendship is not a term commonly associated with these species, or with turtles in general, yet one wonders what a closer examination would reveal.

Not long ago researchers demonstrated that certain garter snakes have an affinity for one another; might we learn that certain turtles do too, finding each other when they emerge from creek-bottom torpor, making space for old acquaintances on the log where sunshine warms their winter-chilled blood? Researchers have even observed how carpenter bees like those who frequent the stormwater pond's beardtongue patch interact inside their nests, which contain both related and unrelated bees. Females are less aggressive toward familiar, unrelated bees than toward relatives they don't know well. Perhaps even bees could be said to have friends.

It's speculation, to be sure, but of a sort that science now makes reasonable and even likely, rather than sentimental projection. And though associations don't automatically imply friendship, they provide fertile ground—or perhaps water, given the context; what of the sunfish swimming below the turtles, able to recognize one another, and living in multigenerational colonies?—for affection. "What would be the setting in which you would have long-term proximity, but not any sort of positive interaction?" asks Smuts. "It's hard to imagine a context in which that would be adaptive."

The possibility opens a different, more intimate perspective on nature than comes from thinking of life purely as a race to propagate one's own genes, with genetic relatedness trumping any other bonds. And although friendship, so profoundly important to our lives, lends itself as an example, the principle is not only about friendship. It's about meaningful social life.

For all but a few species, the cognition and emotion we've explored take form in social settings, as is the case for us. And just as our own social intelligence gives rise to culture, communication, norms of behavior, complex social organization, and group deliberation, the same holds for animals. They are not merely social; they possess society.

◆　◆　◆

IN SPRING AND SUMMER FAMILIES OF WHITE-TAILED deer feed near the paths along Rock Creek. They're beautiful creatures, and though in many suburban areas high deer populations and their appetite for garden plants and forest understories cause some people to see them as pests, they never cease to captivate me.

In Maine, where I grew up, deer would flee at first sight. They're extremely wary there, and rightly so, as people annually kill about one-eighth of the state's population. Along Rock Creek they'll still depart if you get too close but otherwise are untroubled by human presence. Sometimes if you fail to notice a deer, thus leaving your body language natural and unconcerned, they'll allow a much closer approach than when someone self-consciously tries not to disturb them. There's something deeply satisfying about this. Cities are supposed to be unnatural environments, yet here deer treat people as fellow animals.

Their groups usually contain several adult females, a few males, and assorted fawns and yearlings. The females are related: mothers, sisters, daughters. Sons stay with the group until their second autumn, when they depart to find a new group. The young ones learn from their elders the lessons of deer life: how to interact with one another, how to behave around humans, what to eat.

White-tailed deer don't undertake the long-distance journeys of some of their ungulate cousins, such as mule deer, who travel 150 miles between summer and winter lands in the western United States, or caribou, whose round trips cover 750 miles. Urban white-tails also have smaller ranges than their rural counterparts. Still, they move between seasonal territories; they must know where to find grasses and shrubs at different times and the location of especially bountiful patches, as well as where to go and how to return.

Once these movements were believed to be primarily driven by instinct, or at least simple behaviors: deer followed their noses rather than possessing the detailed knowledge of landscapes found in exceptionally intelligent, long-lived animals such as elephants, whose matriarchs lead their herds on ancient routes between seasonal water holes. Research now suggests that learning and memory play a crucial role in all sorts of ungulate migrations. This role is difficult to verify scientifically, but it was demonstrated in a study of moose and bighorn sheep herds relocated to new areas. At first, not knowing the landscape, they stayed in place; as knowledge accumulated over the next several decades, they became migratory, with each generation refining their elders' paths. So it is for these suburban deer carving out a life in parks and median strips and backyards. Knowledge of the land is part of their culture, the body of knowledge that is passed between individuals and persists across time.

Animal culture has been described in many species: an Australian community of bottlenose dolphins who use sponges to flush prey from the seafloor, the greeting rituals of white-faced capuchin monkeys in Costa Rica who insert their fingers into each others' eyes, carrion crows in Japan who drop nuts on roads where passing cars crack them open. At its core is the ability to learn from others. Even bumblebees have, in experimental settings, proved capable of learning to solve puzzles and identify blossoms by watching other bees, raising the question of what information they might share in the wild.

But I find something especially moving about culture and animal migration, the way these extraordinary journeys are guided by lessons shared across generations. It wasn't so long ago that humanity's own migrations were performed this way. Culture is also key to many bird migrations, most famously demonstrated in whooping cranes. After being driven extinct in the wild and saved by captive breeding, birds raised in captivity did not know where to fly; conservationists piloting ultralight aircraft led them over the eastern United States, rekindling a tradition that the cranes later taught to their own progeny.

Most migrating species have not yet been studied this way, but those who have point to how widespread migratory cultures are. As one recent review of animal migration and sociality noted, even birds thought to fly solo "often migrate at the same time and in the same direction as thousands of other migratory birds, providing chances for information exchange." The swifts who roost in the chimney of an apartment building near the stormwater pond arrive from South America on seven-inch-long wings and, perhaps, the wisdom of their people.

The seminal description of animal culture, however, took place in the 1930s in Swaythling, England, where villagers noticed that some of the milk jugs delivered to their doorsteps were open. The culprits were Eurasian blue tits, tiny songbirds belonging to the same genus as chickadees, who had learned to pry caps off the jugs. Soon reports came of jug-opening birds in adjacent towns, and birders meticulously documented the behavior as it spread across the country. Incidents did not happen randomly, as would be expected if birds discovered the techniques on their own, but proceeded stepwise, town by town—a pattern best explained by their learning from each other.

Bottle opening may seem mundane compared to migration, yet in a way it's equally magical, pointing to how culture suffuses everyday life. It's even been suggested that house sparrows and pigeons, the most common urban birds of all, thrive in cities not only because they're resilient or find niches resembling ancestral habitats, but because they pool what they know. Even the most common, everyday birds, doing common, everyday things, have their own bodies of knowledge.

◆　◆　◆

AS IS THE CASE WITH OUR OWN CULTURES, THOSE OF
animals don't only encompass what they know and do but also
how they communicate. Early observations took place around San
Francisco in the 1960s, when biologist Peter Marler discovered
that white-crowned sparrows living in different parts of the region had subtly different songs. At the time, birdsong was thought
to be genetically predetermined. Marler's research showed that
they learned from their parents. He described the neighborhood-
specific variations as dialects, akin to how people living in New
York City and Long Island have distinct accents. Dialects have
since been described in many other species; they're very likely
possessed by the white-throated, song, and house sparrows in my
old neighborhood. A community of the latter nest in ornamental
cement work along the alley beside my apartment building and
gather in shrubs next to the gas station parking lot. It cheers me to
think that even they are unique.

Marler was also interested in the meanings of animal communication. In the early 1980s, using the then novel experimental method of observing how animals responded to recordings
of their calls, he helped show that vervet monkeys had specific
alarms for different predators: one to alert fellow monkeys to
an approaching eagle, another for snakes, another for leopards.
Implicit in the findings was the notion that the monkeys commu-
nicated referentially—that is, their vocalizations referred to some
external object or event, not just an internal state, as does a grunt
of satisfaction—and with intent. It was a conventional wisdom-
challenging assertion. At the time, recounted the *New York Times*
in Marler's obituary, it was widely held "that animal communication, in general, was less like human conversation than like a
bodily function."

Both referentiality and intentionality are hallmarks of
language—and language, argued many scientists and philosophers

then and even now, is unique to humans. It's said to be unlike any form of animal communication and to shape thought in ways no animal can share. The latter claim is at best a hypothesis, but human language is unquestionably remarkable; we coin new words with extraordinary frequency, arrange them according to intricate rules, embed messages within messages, and even represent them visually, as words on this page.

Yet there's a different way of looking at language. The essence of communication is information exchange, which is ubiquitous in the animal world. Language is only one way to accomplish this; much can be done without it. Indeed, many of our own communications are not linguistic at all. A hug, a wave, a smile, a whoop of joy: these are vital to human experience but not unique, and none require syntax, recursion, or any of the high-level properties that define language.

Animals also communicate in forms we don't often think of as rich in information. A few fence posts along the stormwater pond's path are favorite destinations for neighborhood dogs, who usually stop to sniff and to make their own contribution. To a dog's exquisitely sensitive nose—packed, on average, with 300 million scent receptors, compared to our paltry five million—these traces of urine form a detailed record of who visited; when they visited; their age, health, and sexual status; and perhaps their mood. The classic idea of dogs "marking their territories" has been replaced by a far more nuanced understanding. The fence posts are more like canine bulletin boards.

Similar behaviors are found in many other species who use centralized scent deposits as a source of social information. Beavers do so, as do river otters, bears, house mice, and moose; when garter snakes return to their winter hibernaculum after spending summertime apart, they lay a chemical trail for other snakes to follow. Were our noses not more sensitive every landscape would be full of messages.

Gesture is another rich mode of expression. One study of chimpanzees found 36 forms of a single gesture, the meaning of which could change when it was applied to different body parts, used in at least 26 contexts, among them greeting, giving comfort, and sharing food. A squirrel's semaphore-like tail-waving is rather simpler, but there are meanings there, too; even crayfish waving their antennae convey information. And the mere act of touch, as when a crow preens her mate after he squabbles with another crow, communicates something that doesn't need to be put into language

Much as some supposedly solitary creatures prove to be quite social, so there are many whose communications are unexpectedly elaborate. Among them are turtles. In the mid-2000s a graduate student studying northern snake-necked turtles in Australia documented their use of at least 17 different types of vocalizations, from chirps and hoots to howls and drum rolls; aquatic turtles had previously been regarded as largely silent and incapable of complex vocal communications. That discovery inspired yet more studies of turtle vocalizations, such as those used by South American giant river turtles to coordinate migrations and the care of their young. What Rock Creek's red-bellied and painted turtles say to each other on those long lazy afternoons atop their logs isn't known, but findings like these expand our sense of what's possible.

Many animals also possess what might be called language-like systems: they don't have all the properties of human language but do possess some of them. The referential communication described in Marler's monkeys and many species since—the first birds documented to do this were chickens telling flockmates about food—is one such property. Another is syntax, or the ability to change the meaning of sounds by rearranging their order.

Syntax is the subject of an especially ingenious line of research conducted by ethologist Toshitaka Suzuki. As Marler and colleagues did with monkeys, Suzuki recorded the vocalizations of Japanese tits—small songbirds closely related to those bottle-opening Eurasian tits and the pond's black-capped chickadees—then played the sounds back and gauged the birds' reactions. Not only did they communicate referentially—"Snake!"—but when Suzuki reversed the sequence of words conveying "scan for danger and come here," they elicited no response. They possessed syntax, not so different from how a phrase like "open the door" is rendered nonsensical as "door the open." What's more, Suzuki showed that after hearing the word for snake, Japanese tits became alert to anything long and cylindrical. Hearing the word seemed to bring an image to their mind's eye, much as it does with us.

In Suzuki's view, these findings are not relevant only to Japanese tits; he believes they illustrate what may occur in many species. Such studies have not been conducted with black-capped chickadees, but scientists who study them suspect they have similar capacities. Upon closer examination even their namesake chick-a-dee call, which they use in a variety of contexts from keeping flockmates close to warning of predators to announcing the presence of food, proves to be quite complex, varying in harmonic richness, note types, and frequency. What sounds to our coarse ears like one simple call is actually a multitude of forms.

What meanings could be produced within them? Expressions for the weather, for feathers, for features of the natural world that we barely even notice but are important to them, such as the taste of air before a long cold snap or the seed-caching virtues of different types of bark? Ways of saying "How are you feeling?" or "Good to see you!" that, unlike alarm calls and the responses they clearly evoke, are difficult to test experimentally? When one parent returns to the nest with food for their fledglings and the other departs, have they negotiated care duties, as common murres do? Shared information about wind conditions or an insect hatch? It's at least something to imagine when watching a chickadee or most any other social species.

The sharing of information also crosses species boundaries. Much communication is public; many species in the pocket forest respond to the alarm calls of squirrels and nuthatches. Groundhogs listen to the alarm calls of chipmunks. In winter, when feeding flocks—robins, starlings, sparrows, nuthatches, chickadees—gather in the copse beside the pond to eat the fat-rich English ivy berries, it's not uncommon for a single alarm to be followed by complete silence and then, several seconds later, the wing flaps of a hawk's arrival.

The dawn chorus—the phenomenon in which birds sing in unison, at intensities not found at other times of day, shortly before dawn—proffers another example. Early on a spring morning the world below our 17th-floor balcony sounded like a stadium at a distance, positively throbbing with voices. The function of the dawn chorus has been widely debated and no scientific consensus exists, but there's one explanation that makes a great deal of sense: they're exchanging information. The dawn chorus might be understood as a sort of morning newspaper.

Whether or not the birds use language is immaterial. And although it remains true that no animal has yet proved able to master human language, consider a point raised by the linguist Leonie Cornips, who writes that "as far as I know there are no experimental or naturalistic studies showing that humans were successful in acquiring chickens, whales, cows or blue tits' language as a first or second language." For all our intelligence, nobody has yet learned to communicate as chickadees do.

◆   ◆   ◆

THE COGNITIVE DEMANDS OF SOCIAL LIFE ARE CONsiderable. A social animal needs to recognize other individuals—by sight, scent, and sound—and remember what they've done in the past, predict what they might do in the future, and then cultivate relationships, with all the communication and negotiation that entails. Even a song sparrow's life history demands a lot of intellectual firepower, and some scientists have theorized that sociality is the great evolutionary driver of intelligence. Being social demands being smart.

One particularly useful ability is an understanding of what others are thinking. In humans this takes a form called "theory of mind," which means you can reason in complicated ways about what other people know, and even understand that their facts and beliefs differ from your own and might sometimes be false. Whether other animals possess this is a matter of ongoing dispute; great apes do, but evidence for cetaceans and corvids is disputed, and those are cognitive superstars.

It's probably true that human-style theory of mind is a distinctive trait. Yet—once again—this highly specialized ability is composed of simpler cognitive elements that are widely distributed among animals. These include inductive reasoning, empathy, and a sense of what to expect in a given situation; put them together and they produce the ability to attribute mental states to others. When a mother squirrel responds to her baby's cries of distress or a neighborhood crow waits until nobody is watching to hide food atop a telephone pole, they're not necessarily theorizing at length. They simply have a sense of what other animals know and act accordingly. We do this too. If you're out for a walk and see someone scowling and muttering under his breath, you don't need to wonder why. It's enough to know that he's in a bad temper for you to cross the street.

These quick rule-of-mental-thumb assessments may be at the root of anthropomorphism, or the projection of human states of mind onto other animals (or, for that matter, inanimate objects; I can still remember an auto shop sign featuring a sad-looking cartoon muffler that moved me to tears as a child.) Anthropomorphism has a bad reputation among scientists, many of whom have historically associated it with careless sentimentality. But some researchers now argue that it reflects a deep-seated impulse. We project our experiences because for millions of years our ancestors benefited from doing so. We're biologically predisposed to anthropomorphize—and turtles turtle-morphize, deer deer-morphize, and so on.

The philosopher Kristin Andrews sees the ability to attribute mental states to others as a fundament of morality in animals. Theirs is not necessarily a human-style morality, codified in texts and pursued upon ethical reflection, but something simpler: a sense of how one ought to behave. It might be called proto-morality.

It comes from watching what others do, argues Andrews, especially one's mother and later one's peers, whose actions provide a template of social behavior that combines with innate predisposition. Kindness, or prosociality in research argot, is one such predisposition, observable in the tendency of rats to help distressed peers or the way mother raccoons adopt orphans. Another is a sense of fairness, or what scientists call inequity aversion. A dog who performs a trick when asked may quit performing upon seeing another dog rewarded for that same behavior; the disparity rankles. Crows and ravens also seem to have a similar aversion to inequity, underscoring that this behavior isn't exclusive to mammals, but is something shaped by social pressures across the animal kingdom.

Play provides a valuable opportunity for animals to develop their moral sensibilities. "While individuals are having fun in a relatively safe environment," writes ethologist Marc Bekoff, "they learn the ground rules about what behavior patterns are acceptable to others." If you pay close attention at a dog park, you'll notice how they'll take turns while wrestling, with big dogs allowing little ones to win; when one behaves too roughly, gestures of reconciliation follow, and a dog who doesn't follow these codes of conduct is soon excluded. Rats behave much the same way.

Animal proto-moralities won't necessarily resemble our own, but they're social norms nonetheless. Mix them with relationships, culture, and communication, and the foundation is laid for *society*—something that's appreciated when it comes to certain animals, such as African elephants with their matriarchal hierarchies, the alliances of chimpanzees, and orca clans who actually line up to greet one another in formal ceremonies. Yet those examples create what is, to my mind, a too-narrow view, with society limited to a few extra-special species. When garter snakes reunite in winter, black-capped chickadees disperse to breed as pairs and then rejoin their flock in autumn, or bluegill sunfish gather at their spawning colonies in Rock Creek, those too are societies of a sort. Nature is full of them.

◆　◆　◆

AS DUSK FALLS OVER THE POND, THE FLITTING SHAPES of big brown bats are visible in the half-light. Come dawn they'll roost in a big tree somewhere nearby—an old oak with a hollow trunk, perhaps, or a shagbark hickory with peeling bark forming crevices perfect for resting. Big brown bats are so inconspicuous that the tree could be in someone's backyard without the people living there even noticing them. A whole society might be hidden above their heads.

In recent years scientists have discovered fascinating things about the lives of many bat species. Sac-winged bat mothers vocalize to their pups with exaggerated, high-pitched intonations reminiscent of what we call motherese; the pups themselves babble, much like human babies. Vampire bats regurgitate meals in order to feed their comrades, creating a "biological market" where favors are exchanged and values fluctuate according to complicated rules of reciprocity. Fruit bat vocalizations—not their ultrasound prey-detecting sonar, but interpersonal squeaks—are extremely complex and used to deliberate roost sites, food sharing, and of course sex.

I asked Mark Brigham, a biologist who studies big brown bats, whether their social lives might be similarly rich. That has not been formally studied, Brigham says, but he suspects it's the case. Big brown bat communication systems are quite sophisticated; they don't trade food but might be trading information. Related females found new colonies, but on any given night, as its members decide which of several trees to spend the day in, they break into smaller clusters of unrelated individuals, some of whom have long-term affinities. These could be friends, Brigham says, though he's not comfortable with the term.

Big brown bat society is what scientists call fission–fusion society, with groups forming and dispersing and coming together again. It's a form shared by, among many other species, our own. How the bats decide on roost sites or whether to join a new colony is unknown, but a deep vein of research describes the many ways animal groups make decisions.

Only sometimes does a single dominant animal decide for everyone. Honeybees take days to decide whether to found a new colony after a queen is born; winter flocks of black-capped chicka-dees seem to deliberate about when to start foraging in the morn-ing; and the mathematical signatures of pigeon flock movements suggest they're actively negotiating while in flight. Some species even vote. Red deer groups move only when two-thirds of a herd agrees on a direction; American bison evidently require a major-ity but tend to follow the lead of older females. Among hamadryas baboons, a majority of adult males rule the day, and votes are indi-cated by shifting position on rock seats. African wild dogs vote by sneezing, with three supporting sneezes sufficient for a pack to fol-low a high-ranking dog's request to move, while a low-ranking dog requires ten sneezes of approval.

There's a good evolutionary reason for all this. Consensus decision-making systems, and in particular forms of democ-racy, allow individuals with different information to share what they know. The group benefits by considering every perspective. There's an even deeper significance, though: these animals are not just engaging in rote individual behaviors that produce collective action but thinking things through. They're demonstrating them-selves to be not only a group or a population, but a true community.

My favorite example comes from whooper swans, who signal their preference for departure by honking and leave only when the frequency of honks crosses a certain threshold. I asked Jeffrey Black, one of the biologists who discovered this vote-by-honking system, whether Canada geese might share it. Black thinks they do. They may also send signals by flashing the white patches on their cheeks, he added.

In the evening's last light, the flitting shadows of big brown bats hint at social worlds of which we have only a glimpse, but that are perhaps no less meaningful than our own. Sometimes their chirping calls can be heard above the ventilation system hum of nearby apartment buildings. Canada geese bob in the water, dimly illuminated by lamps that have flickered on around the pond's walking path. One honks, then another, and another, until it reaches a crescendo and they rise into the gloaming, black silhouettes against a deep blue sky. Their fading calls are the sound of a people.

II

RETHINKING
ANIMALS

4

A New Nature Ethic

IN THE FRENCH PAINTER ÉMILE-ÉDOUARD MOUCHY'S 1832 *A physiological demonstration with vivisection of a dog*, thirteen men gather around a dog tied leg and neck to a table. The room is dim, lit only by a single overhead window that shines brightest on a white-smocked man with rolled-up sleeves who slices open the dog's chest with a scalpel. The brown-and-white coated spaniel pulls helplessly against the cords, mouth agape, snapping, and howling in pain. Blood soaks her fur. The men are a study in clinical attentiveness, taking notes and conversing, peering intently into the wound. Another dog leashed to the floor struggles to reach his pinioned kin.

It is unclear how Mouchy himself felt about this grotesque tableau. Perhaps he was horrified. Or, more likely, as suggested by the lighting, he might have considered it a literal vision of Enlightenment progress. Either way it's a faithful recording of a scene common in the Age of Reason, when scientists and doctors learned physiology through living dissections that could take days to com-

plete. "They administered beatings to dogs with perfect indifference, and made fun of those who pitied the creatures as if they had
felt pain," wrote one observer. "They said the animals were clocks;
that the cries they emitted when struck were only the noise of a
little spring which had been touched."

The observer was a secretary of the Jansenist monkish order,
part of a Catholic movement associated with René Descartes, the
seventeenth-century philosopher, mathematician, and scientist
sometimes described as the father of modern philosophy. Descartes famously likened animals to mechanical automata: their
apparent consciousness was an illusion, as only humans possessed
souls and therefore minds. Other creatures were essentially exquisitely crafted wind-up dolls. Absurd as that sounds to modern ears,
it drew upon a well-established belief system.

Long before Descartes, the Greek philosopher Aristotle, arguably the most influential thinker in Western history, turned his
attention to animals. He described Mediterranean fauna in exquisite detail; his systematic classification scheme presaged modern
taxonomy, and in some ways he appreciated animal intelligence.
Aristotle's zoological works contain passages remarking on their
thoughtfulness: octopi concealing themselves in clouds of ink did
so knowingly, certain wild goats treated arrow wounds with herbs,
dolphins tended to a dead baby as if moved by pity, and bird communications compared to speech. He put humans on a continuum
with animals, differing in degree rather than kind.

At the same time, this continuum existed on a *scala naturae*, or
"ladder of nature," a hierarchy not only taxonomic but moral, with
rational beings—humans—perched at the top. In his political writings, Aristotle described animals as incapable of rational thought.
He limited the scope of their experiences to hunger and pain.
Lacking reason and speech, of course, it followed that animals did
not belong to the polis, his hallowed political community. Those
bird communications were so much noisemaking. In that Aristotle was truly foundational. Like generations of aspens growing

from the same roots, this argument recurs through the next two thousand years of Western thought: a categorical barrier separates humans from all other animals, and that barrier is grounded in some cognitive trait whose absence is rationale for diminished moral significance.

If Aristotle was at least conflicted about animals, though, the Stoics who followed in his wake were unambiguous. They believed that animals could not think, and they made rationality the basis not only of political membership, but of entitlement to Earth. "Suppose someone asks for whose sake this vast edifice"—the world—"has been constructed," asked the Stoic philosopher Balbus. "For the trees and plants, which although not sentient are sustained by nature? No, that is absurd. For the animals? No, it is no more plausible that the gods should have done all this work for the sake of dumb ignorant animals. Then for whose sake will anyone say that the world has been created? Presumably for those animate creatures which use reason: that is, for gods and men."

In contrast to the Stoics, the Epicureans—whose name survives today as an adjective describing sophisticated taste in food and drink—made sensual pleasure, not reason, their great virtue. But they too shared a disdain for animals, adding their own wrinkle to the idea of reason as a distinguishing human characteristic. Animals "do not have the power of making a covenant to not harm one another or be harmed," argued Epicurus, so for them "there is neither justice nor injustice." In other words, morality resides in formal contracts, which animals cannot make; thus they have no morality and are undeserving of our own moral regard.

Not every Greek thinker felt this way, however. In *On the Cleverness of Animals* and *Beasts Are Rational*, Plutarch acknowledged the "purpose and preparation and memory and emotions" of animals and also their "courage and sociability and continence and magnanimity." He actually considered animals to be, in a sense, *more* rational than humans because they were not tempted by luxury or excess. Several centuries later, Porphyry denounced those who elevated humans by denigrating animals, which he likened to denying that one bird could fly at all simply because another flew even higher. "Since, however, justice pertains to rational beings, as our opponents say," chided Porphyry, "how is it possible not to admit, that we should also act justly towards brutes?"

Such views would not prevail but are important to note. The path to understanding animal minds has not been straightforward, beginning with mistaken but understandable denial and ending in contemporary recognition. Instead, arguments over their capacities have occurred for millennia, with ancient misconceptions persisting from epoch to epoch. They have been inextricable from how people relate to animals: as objects to be exploited or as kin to treat ethically.

The scientific and scholarly traditions that triumphed were those that proceeded as though only human intelligence was meaningful. Science now makes that belief untenable, but its consequences won't be neatly erased by new findings. The denial of animal intelligence helped lay the foundations of our society, shaping ideas about animals—and, ultimately, our relationships to them, and to all of nature—well into the present. Those cultural legacies remain even as their premises erode, and reckoning with them can help point us toward different ways of being.

◆ ◆ ◆

THOUGH ARISTOTLE PREDATED CHRIST BY SEVERAL centuries, his *scala naturae* would be adopted by medieval Christian thinkers, who simply added angels and God above humans in what they called the Great Chain of Being. They also carried on the tradition of emphasizing differences between humans and animals, which historian Joyce Salisbury attributes to the desire of early Christian thinkers to distinguish themselves from paganism.

"They repudiated many classical beliefs, including attitudes toward sexuality, entertainment, bathing, and other social and cultural practices," Salisbury wrote. "In the process, they also rejected a classical view that saw humans and animals as closely related." After all, only man was created in God's own image.

Thirteenth-century philosopher Albert the Great asserted that animals had sperm unaffected by heavenly bodies. Quadrupeds, he wrote, walk on four legs because their "innate heat" is "inadequate to maintain them in an erect posture." Evidently bipedalism was a function of body temperature. As for mental differences, animals' perceived lack of rationality was again their defining quality. Theologians interpreted complex behavior entirely in terms of instinct: A sheep running from a wolf, for example, exhibited no more awareness than someone jerking their hand from a flame. Saint Thomas Aquinas, a student of Albert whose teachings shaped Catholic thought, wrote that animals are "without intellect" and that their instinctive reactions are "as inevitable as the upward motion of fire." Fascinated by mechanical clocks, Aquinas likened animals to machines, striking the note later made infamous by Descartes.

Medieval philosophers' belief in animal mindlessness and, by definition, lack of an immortal soul, "formed the basis for their belief in the 'natural' dominion of humans over animals," wrote Salisbury, and indeed over nature and all its communities. "Let the human race recover that right over nature which belongs to it by divine bequest," wrote Francis Bacon, the philosopher and statesman who devised the scientific method and, along with Galileo, is considered a founder of modern science. Balbus and the Stoics would have been proud.

The Scientific Revolution and the Enlightenment displaced Christianity at the heart of Western knowledge, but theological attitudes toward animals and nature simply evolved, now buttressed by the seeming objectivity and rigor of science. Descartes' likening of animals to mechanical automata embodied his belief in the separation of matter and soul: the actions of animals' bodies had nothing to do with their minds. And because animals lacked reason, their minds effectively contained nothing, not even awareness of their own pain, hunger, or thirst. As evidence, Descartes presented animals' lack of language. "There are no men so dull-witted and stupid, not even imbeciles, who are incapable of making their thoughts understood," wrote Descartes, but no animal possibly could—which "shows not only that animals have less reason than men, but that they have none at all."

Other now-canonical thinkers would build on that categorical differentiation of humans and animals. Among them were Thomas Hobbes, architect of the social contract, who carried forth Epicurus' argument that animals could not enter formal agreements and thus had no rights; John Locke, who held that while certain human rights were inalienable, the same did not apply to animals; and Immanuel Kant, from whom we inherit the principle that every human being is intrinsically valuable and should not be treated as means to someone else's ends. But as for animals, argued Kant, a man could treat them "as means and instruments to be used at will for the attainment of whatever ends he pleased."

Kant did encourage people to be grateful for their labors and to kill animals only when necessary and with mercy, though not because animals themselves deserved any regard. Rather, doing otherwise demeaned human virtue. "Thus only for practice," lamented the philosopher Arthur Schopenhauer, a contemporary of Kant, "are we to have sympathy for animals." Schopenhauer was one of many Enlightenment philosophers who, like Plutarch against Aristotle, pushed back. He agreed that animals lacked reason—but to him, that wasn't very important. They could suffer and therefore deserved our compassion. Jeremy Bentham, the founder of modern utilitarianism, captured this in a quote that's often repeated today: "The question is not, Can they *reason*? nor, Can they *talk*? but, Can they *suffer*?"

Notable Enlightenment animal sympathizers also included Thomas Paine, John Stuart Mill, and Voltaire, who considered it nonsensical that animals should possess the physiology underlying human feelings but not the feelings themselves. The back-and-forth underscores two things: disregarding animal intelligence was not an intellectual inevitability but a choice, and it was inseparable from a vision that—as ecologists Frédéric Ducarme and Denis Couvet write in their history of how nature is understood—left the natural world "entirely open for appropriation and exploitation." It was a belief system perfectly suited to a colonialism dependent not only the brutal erasure of human societies and exploitation of human lives, but those of nonhumans as well.

Alternative traditions of knowledge and relations to animals did exist, of course. Though the diversity of precolonial cultures can't be distilled into a monolith, it's fair to say that a sense of animals as sharing basic mental properties with humans, of humanity as having moral obligations to them, and of humans as a part of the world rather than its metaphysical center, was common. As views of mindless animals and nature as resource triumphed in the colonial world, however, Indigenous value systems were literally exterminated. Meanwhile even positive colonial relations to nature reflected their intellectual heritage.

A telling anecdote comes from John James Audubon, the great artist, ornithologist, and namesake of the Audubon Society, who wrote of a morning spent with a farmer who had trapped three wolves in a pit. The wolves "were lying flat on the earth, with their ears close down to their heads, their eyes indicating more fear than anger," recounted Audubon. They did not struggle as the farmer leaped into the pit, cut their hamstrings, and bound them with rope. Audubon, who had been "glad of the opportunity," helped the farmer hoist the first wolf from the pit.

The wolf was "motionless with fright, as if dead, its disabled legs swinging to and fro, its jaws wide open, and the gurgle in its throat alone indicating that it was alive." The wolf did not fight as the farmer dropped him onto the ground for his own dogs to tear apart. The second wolf did not fight, either. But the third wolf "shewed some spirit" and "scuffled along on its forelegs at a surprising rate, giving a snap every now and then," managing to bite a dog before the farmer shot the wolf. His dogs then "satiated their vengeance."

Wolves had killed many of the farmer's sheep and one of his horses. His desire to kill them was understandable. The brutality of his methods, though, and the pleasure Audubon took as a spectator, is unsettling. The episode embodied a relationship to nature in which certain animals might be appreciated and even treasured—but others could be treated as a morning's blood sport. No compassion nor ethical consistency nor even a second thought was merited. Audubon was a nature-lover and early conservationist, but he was raised in a world of normalized domination.

◆ ◆ ◆

CHARLES DARWIN IS ALL THE MORE REMARKABLE IN contrast to his time. Steeped in the dogma of categorical human uniqueness, he described animals with empathy—"our fellow brethren in pain, diseases, death, suffering and famine—our slaves in the most laborious works, our companions in our amusements"—and traced their and our origins to a common ancestry. He followed the implications of evolution, arguing that "most of the more complex emotions are common to higher animals and ourselves" and that "difference is in degree, not in the nature between animals' and humans' way of reasoning."

Darwin's work presaged a late-nineteenth-century surge in the regard of animals as fellow thinking, feeling beings. His own protégé, George Romanes, dedicated his career to studying their mental lives. But today, Romanes has a reputation for being careless, accepting anecdotes without question, such as a secondhand just-so story about a monkey who proffered a bloody paw in order to shame a hunter. Such uncritical credulity, the standard narrative holds, threatened to throw the study of animal behavior into disrepute.

A case can be made that Romanes and his peers were far more scientifically rigorous than they're now given credit for, and they were unlucky victims of an understandable backlash against the popularity of occultism and supernatural thinking within that era's psychological science. In any event many scientists revolted; a new paradigm emerged. Romanes' own protege, Lloyd Morgan, offered guidance. "In no case is an animal activity to be interpreted in terms of higher psychological processes," advised Morgan, "if it can be fairly interpreted in terms of processes which stand lower in the scale of psychological evolution and development."

In other words: assume the animal is an automaton. Scientists were expected to explain behavior in terms of instinct and thoughtless response until every possible alternative was conclusively eliminated. The principle was so influential that it's now known as Morgan's canon, and though it didn't technically deny animal intelligence, but simply required claims to be backed by exhaustive evidence, in practice it raised the barrier anew. There followed a century of science that again regarded animals mostly as machines, an attitude neatly expressed by the pioneering behaviorist B.F. Skinner, who described animals as "conscious in the sense of being under stimulus control." By the late 1970s, when Donald Griffin's *The Question of Animal Awareness* posited that animals could think and reason, it was considered a radical statement in scientific circles.

Around the same time that scientists who believed in meaningful animal intelligence were being castigated, something similar happened to popular naturalists. At the turn of the twentieth century, public interest in nature had surged, and naturalists were a prominent part of public life. Among the best-known were Boy Scout cofounder Ernest Thompson Seton, who wrote that "animals are creatures with wants and feelings differing in degree only from our own"; Canadian poet Charles Roberts; and William Long, who described animals as individuals and in communities, thoughtful and emotional, relying upon reason and learning. Their

books were best-sellers, with biographies of animals that emphasized kinship of animals replacing scientific treatises and evolutionary epics as popular fare.

Yet some of these naturalists' claims, such as Long's tale of
a woodcock setting a broken leg and wrapping it in a mud cast,
strained belief. They earned the enmity of naturalist John Burroughs, who described Long and company as "yellow journalists
of the woods," and then Burroughs' close friend President Theodore Roosevelt, who publicly denounced them as "nature fakers."
Though little remembered now, the so-called nature fakers controversy was front-page news—the *New York Times* called it "The
War of the Naturalists"—with the camps trading rhetorical blows
for years. At stake was not only an understanding of animal minds,
but a worldview. As historian Ralph Lutts recounts, the surge of
interest in nature occurred as an urbanizing, post-frontier society
reconsidered its relationships to nonhumans. "Nature lovers were
empathizing with their 'poor relations,' rethinking their ethical
and ecological relationships with the natural world, and acting to
protect wildlife," writes Lutts. Increasingly, they embraced a view
in which humans and animals "shared the same moral ground and
similar emotional and mental lives."

In that context, criticisms of the nature fakers also appear as
defensiveness to a new moral order. Burroughs himself was clear
about that. In an essay entitled "Machines in Fur and Feathers,"
he admitted that he "could not be a party, directly or indirectly, to
the murder of beings between whom and myself there existed the
relationship implied by the gift of the faculty of reason." Fortunately, wrote Burroughs, there is "very rarely anything that supports" that view. As for Roosevelt, an avid recreational hunter who
on a single African expedition killed no fewer than 11,400 animals,
and of whom Long wrote that "every time he gets near the heart of
a wild thing he invariably puts a bullet through it," the enmity was
undoubtedly personal.

Burroughs, Roosevelt, and their allies—among them Frederic Lucas, later the director of the American Museum of Natural History, and C. Hart Merriam, head of the US Biological Survey—ultimately prevailed. Their opponents continued to write, but they were relegated to genre fiction. In the decades that followed, wild animals became the formal domain of wildlife managers focused on hunting, fishing, and trapping; behaviorism's view of animals as stimulus-response machines dominated the sciences; and public attitudes toward nature formed in a milieu where relating to animals as fellow thinking, feeling beings was not encouraged.

That kinship of mind and body between humans and animals had, for a time, been a widely held belief would be mostly forgotten. Consider the modern understanding of environmentalist John Muir, the first president of the Sierra Club, who described the "humanity" of the animals he'd known on his childhood farm—a sympathy at odds with the "mean, blinding, loveless doctrine" that "animals have neither mind nor soul, have no rights we were bound to respect, and were made only for man." He took this attitude into his beloved wilderness, writing of animals as "fellow citizens," "feathered people," and "horizontal brothers," and scathingly condemned recreational hunting as "murder."

"When are you going to get beyond the boyishness of killing things?" he reproached Roosevelt during their famous camping trip in Yosemite, which is credited with inspiring Roosevelt to create America's national park system. Apart from being lionized for wilderness preservation, Muir is now remembered as a bearded ascetic who found God expressed in nature's beauty.* As for his

---

* Muir's attitudes toward Indigenous people are a subject of considerable controversy. To his detractors, Muir was a racist whose romanticized vision of wilderness was predicated on unacknowledged genocide—a depiction that, say his defenders, ignores the fullness of Muir's views and their growth during his life. This argument can neither be ignored nor adjudicated here, but the counsel of Ben Jealous, former president of the National Association for the Advancement of Colored People and current executive director of the Sierra Club, seems wise. People should be judged at their best, says Jealous, not at their worst.

views on animals, they are seldom recalled. In 2006 the Sierra Club held a writing contest in which the winner received a hunting trip to Alaska.

◆   ◆   ◆

As THE TWENTIETH CENTURY PROGRESSED, A SENSE of animals as intelligent beings was not so much extinguished as unevenly distributed. In some contexts, it continued to flourish: witness the transition to keeping dogs and cats as companions rather than guards or rat catchers, which has likely done more than anything else to open people's minds to animals. Thoughtful animals also featured prominently in popular culture, from children's tales to fare such as *Lassie* and *Dr. Dolittle* to such classics as *Never Cry Wolf.*

Within formal discourse, however, a limited view of animals predominated. Visions of nature that emphasized its beauty, transcendence, and spiritual values did not engage with them as, in Muir's words, fellow citizens. Large-scale landscape protection, from the national park system to the multiple-use stewardship of agencies such as the US Forest Service, presented nature as a scenic retreat or resource stockpile. Wildlife management treated animals as interchangeable species units.

New laws, such as the Lacey Act of 1900 and the Migratory Bird Treaty Act of 1918, did protect some animals, but they represented a species-level recognition rather than individual regard. One could see animals as furred and feathered machines while wanting to preserve their existence. The Marine Mammal Protection Act and, most powerfully, the Endangered Species Act continued this legacy; they were part of a midcentury rise of environmentalism that, while profoundly important and responsible for preserving many species from destruction, found it easier to personify Mother Nature than a mother mallard duck.

A deeper modern engagement with animal minds took place

in the burgeoning animal rights movement, of which Peter Singer and Tom Regan are the best-known proponents. Singer's 1975 *Animal Liberation* was a landmark. He brought the subject of animals into the realm of modern philosophy and considered the capacity for thought and feeling to be more important than a species label; he insisted that animal pleasure and pain be accounted in ethical tallies of benefit and harm. Regan's *The Case for Animal Rights*, published in 1983, took a different tack from Singer's utilitarianism; he emphasized the principle of moral regard, which he believed extended to any creature possessing "perception, memory, desire, belief, self-consciousness, intention, [and] a sense of the future." Any such creature, opined Regan, could not be treated as a means to an end but deserved to be regarded as intrinsically valuable ends in themselves.

Both Regan and Singer had their blind spots, particularly when it came to biological communities—species, populations, ecosystems—which their philosophies excluded from consideration because such collectives are not themselves sentient. More broad minded was the late British philosopher Mary Midgley, whose *Animals and Why They Matter* was also published in 1983. Though less known than Singer and Regan, Midgley's importance has grown steadily; to her, both individuals and collectives had moral claims, though she was less concerned with laying down moral codes than helping people think clearly about them.

She chastised philosophers for having separated themselves from animals—not just conceptually, but for literally failing to spend time with them. If anything makes humans unique, argued Midgley, it's the deep bonds we feel with animals, which she considered as fundamental to human experience as song or dance. Midgley was fascinated by Jane Goodall's observations of chimpanzee society and the work of ethologists like Konrad Lorenz, whose goose research was mentioned earlier. They had started to erode behaviorism's dominance, and this new information about animals, Midgley wrote, "inevitably gives us a sense of fellowship with them." She also critiqued the glorification of rationality and colonial mentalities of the domination of nature.

Midgley articulated the concept of a "mixed community," acknowledging that human lives are intertwined with those of animals, plants, and ecosystems, demanding a richer ethics that recognized each of their claims. Doing so might be difficult, but that was no reason not to try. Assisting in this task was something not often discussed by philosophers. "Love, like compassion, is not a rare fluid to be economized, but a capacity which grows by use," she advised.

In the years since Midgley, Regan, and Singer published their master works, the field of animal ethics has flourished in tandem with scientific investigations of their minds. The humanities have taken what's called "the animal turn," as geographers, anthropologists, sociologists, and historians focus their attention upon human relationships with animals and ask what it means to consider their perspectives. These works have sparked small revolutions in how people think about animals used for food, research, and companionship. Progress is uneven, but conversations are happening, and people are grappling with new ideas.

When it comes to nature, though, a richer reckoning with animal intelligence has been slow to percolate. Even modern conservation, the culture and practice that in many ways defines our relations to wild animals, rests on the scientific discipline of conservation biology, which at its founding in the early 1980s explicitly excluded animal welfare and animal sentience. Moral regard for individual animals was seen as conflicting with ecological processes in which predation and death are essential, so it was better to avoid the subject. Populations and species were the proper scale of focus. "The ethical imperative to conserve species diversity is distinct from any societal norms about the value or the welfare of individual animals or plants," wrote the late Michael Soulé in his landmark 1985 paper "What Is Conservation Biology?"

Soulé did not deny the importance of individuals. Two years before his death in 2020, he and I discussed a push by some conservationists to embrace developments in animal ethics. During our chat, he spoke of the importance of compassion and reminisced about caring for a seagull he found injured on a beach. His voice caught as he recalled that seagull's distress and the feeling of leaving him behind, broken-winged on the sand, to a slow death. Yet still he insisted that considering individual animals—acknowledging their intelligence, wrestling with the ethical questions that raises—was outside conservation's purview.

That tendency is reflected in ideas that have rippled through conservation and environmentalism in the past decade: the imperative of accounting for human as well as environmental well-being, grappling with legacies of racism and inequality, considering the social dynamics that shaped ideals of pristine wilderness, and embracing nature in cities and places altered by human activity. Even the burgeoning "rights of nature" movement, which seeks recognition of ecosystems as legal persons, such that a forest or a river can be understood as having rights violated by their destruction, shies away from rights for animals.

Although theorists now acknowledge that nature is, in a sense, a social construction, shaped by cultural attitudes and historical legacies, the construction of animals goes largely unexamined. This does not seem intentional or mean-spirited; conservation is full of people who love animals. Rather, it's as though avoiding animal minds has become so normalized as to preclude other ways of thinking.

◆　◆　◆

THE STREAM THAT FEEDS THE STORMWATER POND flows from a large concrete drainage channel, just a gentle burble except when it rains. At twilight I would sometimes catch a glimpse of a raccoon mother and her cubs emerging.

I can still picture them in my mind's eye: the mother's shape solidifying in the darkness of the pipe's mouth, pausing to survey her surroundings. After a few moments the youngsters join her. No doubt she'd guided them through the network of pipes and drains with entrances throughout the neighborhood, providing safe passage beneath the streets. Mom then leaves the safety of the pipe and walks into the low bankside vegetation, popping up a way further, standing on her hind legs and sniffing. By some undetectable signal or understanding, her cubs then scamper to her. They make their way to the pond where cattails teem with food and the surrounding fence provides safety. Later, when it's too dark to see, I'll hear the cubs chasing each other through the branches of a small pine. Their feet leave no prints on the concrete embankment, but elsewhere the mud records their front paw prints, which look almost exactly like human hands.

Raccoons occupy a special place in North America's cultural history. In precolonial lore, stories emphasize their cleverness and curiosity: their uncanny ability to pick fruits at the peak of ripeness, the nimble fingers with which they untie knots. They cross boundaries, being equally at home in both deep wilderness and urban edges—and in the early twentieth century, after several colonial centuries of being hunted for their flesh and fur, and being persecuted as agricultural pests, they enjoyed a moment of pop-culture acclaim. Newspapers recounted their antics and many people kept them as pets, in a period when keeping pets itself was an institution in transition, less a matter of utility than affection. Among them was President Calvin Coolidge, whose raccoon Rebecca accompanied him on daily walks around the White House.

For a brief time, they were subjects of study by experimental psychologists, especially by Lawrence Cole of the University of Oklahoma and Clark University's Herbert Burnham Davis. Their experiments on raccoon learning and memory showed raccoons capable of complex reasoning and insight; even as the nature fakers controversy raged, these scientists described their capacity for ideas and human-like thought. That research fell from favor, partly under the weight of growing behaviorist dogma and the sheer difficulty of using rambunctious raccoons, rather than rats or pigeons, as study subjects, though their public popularity remained.

Back in my neighborhood, the mother raccoon stands, her dark black eyes reflecting the twilight sky for a moment, turning them a luminous blue. I'm reminded of an Aldo Leopold essay, "Thinking Like a Mountain," contained in his 1949 *A Sand County Almanac,* one of the most influential works of environmental ethics. It describes how he and a partner encountered a family of wolves playing with their mother on a riverbank. "In a second we were pumping lead into the pack," recalled Leopold. "I was young then, and full of trigger-itch; I thought that because fewer wolves meant more deer, that no wolves would mean hunters' paradise." They hit the mother wolf and reached her "in time to watch a fierce green fire dying in her eyes." After that, wrote Leopold, "I sensed that neither the wolf nor the mountain agreed with such a view." It marked the beginning of his appreciation that wolves played an ecological role in regulating deer populations whose foraging would otherwise denude habitats and cause their own starvation.

Decades before, Leopold wrote about encountering raccoons on a hunting trip down the Colorado River. One, inadvertently caught in a bobcat trap, "looked very wet and lonesome, but his hide looked fair so we skinned him out and reset." Later his hunting partner saw a group of animals along the shore. "We hoped they were cats but as we stole up in the canoe we saw they were a whole family of coons," wrote Leopold. "I took the shotgun and killed one dead with the first barrel and thought I'd killed the big one with the second." The "big one" was only crippled, though. Leopold found him hiding under a pile of dead brush and finished the job. No ethical revelation followed.

That account, contained in the anthology *Round River,* was written in 1922, when Leopold was 35 years old, but published posthumously in 1953, eventually earning the condemnation of Rachel

Carson. "Until we have courage to recognize cruelty for what it is—whether its victim is human or animal—we cannot expect things to be much better in the world," she wrote, and the young Leopold certainly deserved that rebuke, though less so the person he grew up to be. In a *Sand County Almanac* he wrote that "men are only fellow-voyagers with other creatures on the odyssey of evolution," a knowledge that should give us "a sense of kinship with fellow-creatures; a wish to live and let live." His body of thought, known as the land ethic, accorded direct moral consideration to animals—a powerful vision, especially when couched within an ecological awareness that in Leopold's era was truly visionary.

Nevertheless, Leopold's vision of *Homo sapiens* as "plain member and citizen" of nature was not one in which fellow animals appeared as individuals, with societies of their own, nor would this plain member and citizen be expected to forego complete mastery. Raccoons who lived on the once-blighted land that Leopold toiled for decades to restore were still called "seed stock." Though he did acknowledge the subjectivity of animals, down to the field mouse who is a "sober citizen who knows that grass grows in order that mice may store it underground as an underground haystack," it was enough to account for their interests in the population-level and ecosystem-level terms that characterized his vision.

"A thing is right if it tends to preserve the integrity, stability and beauty of the biotic community. It is wrong when it tends to do otherwise," he famously wrote, a marvelous prescription of ecological awareness that influenced generations of nature lovers. He noticed the lone geese who arrived in spring and the "disconsolate tone" of their honking, which might be attributed to their being "broken-hearted widowers, or mothers hunting lost children" shot by hunters lining the wetlands through which they flew the previous fall—but this would not stop Leopold from joining their ranks.

All of this is not to criticize Leopold for being a hunter. It's hard to make peace with how he killed animals for sport, but for me that's outweighed by the legacy of his land ethic and the example of his stewardship. To pioneer the science of game management at a time when hunting was poorly regulated and to conceive of wild animals as belonging to everyone—not just a few men of wealth—and meriting management with ecology in mind, were great accomplishments. It's simply to say that his idea of kinship still had its omissions.

Leopold didn't seem to think very hard about certain things, much less have to engage with the scholarship and social changes of the late twentieth and early twenty-first centuries, which have reconfigured human relationships to animals in profound ways. He didn't need to reckon with scientific research that made clear how deep that kinship is between humans and animals, with critiques of how colonialism is alive and well in our treatment of nonhumans, or with new ideas about giving them legal rights or political representation or taking their interests into account in the myriad ways our lives entangle with theirs.

In this, Leopold is emblematic. He was a thoughtful and nature-loving person, but he was raised in the shadow of that millennia-long saga, from Aristotle to Albert the Great to Descartes to the nature faker correctives, of ignoring animal intelligence and using the differences between them and us to justify denying basic moral consideration. His contribution was enormous but incomplete. Where animals fit into our ethics of nature—not as species, populations, or symbols, but as fellow persons who happen not to be human—is an unfinished project. That task falls to us.

It does so at the dawn of an epoch sometimes called the Anthropocene, in which human influence is literally shaping the Earth, although the implications of that term are debated: does it merely signify our vast power, or call on us to reflect, to adopt ethics that are not only sustainable but just, even kind? The latter, I think, if we follow our conscience, and if Earth is to be a planet worth living on. And given what we know about animals' minds and lives, a relationship to nature that omits those insights doesn't seem fair.

This isn't about declaring hunting and trapping to be wrong,

insisting that all animals should have rights like ours, or conduct-
ing some other seductively simple exercise. It's about asking questions and using our imagination; about thinking through what it
could mean to consider wild animals to be fellow people. The blue
glow in the mother raccoon's eyes, like the green fire that so moved
a young Leopold to reconsider wolves, might be understood as
an invitation to reconsider how we think about wild animals and
about nature.

5

Animal Personhood

WHAT IS A PERSON? AN ANIMAL ADVOCATE MIGHT say: a some*one*, not a some*thing*. To Immanuel Kant, a person was someone who deserved to be treated as an end in themselves, rather than used as a means to achieve another's ends. He reserved this status for "rational beings," by which he meant humans—and if you look up "person" in the *Oxford English Dictionary*, the first definition is "an individual human being."

Yet if you look to *Black's Law Dictionary*, the world's most widely cited legal text, to which judges and lawyers turn when seeking the precise legal meaning of a term, you will find another definition. "So far as legal theory is concerned," states the 11th and current edition, "a person is any being whom the law regards as capable of rights or duties." This reflects a recent change. Previously it read "capable of rights and duties"—a small but, as we shall see, potentially profound difference.

The party responsible for that change from *and* to *or* is the Nonhuman Rights Project, an advocacy group founded in 2007 by attorney Steven Wise. He has practiced animal law since the early 1980s, a time when most US states considered cruelty to animals a misdemeanor akin to public intoxication. Laws have changed quite a bit since then—every state now regards animal cruelty as a felony offense—but they remain flawed in ways that might be surprising. If someone kills another person's beloved dog, for example, they can typically be sued only for the cost of buying another. In the law's eyes, dogs are as fungible as televisions.

That's for the animals we hold in highest esteem, as veritable nonhuman family members. For less-valued creatures, laws are even more of a mess. In the United States, the federal Animal Welfare Act exempts farm animals and most lab animals; the Humane Slaughter Act doesn't apply to chickens or fish, who account for the vast majority of farmed animals. State laws are a patchwork, with loopholes that permit practices most people consider cruel, such as keeping pigs in gestation crates or shooting pigeons for fun.

As for wild animals, meaningful protections are mostly limited to endangered or threatened species. The rest can generally be killed or displaced with minimal fuss. This can be perfectly legal, as when the Virginia Department of Transportation destroyed the state's largest seabird breeding colony as part of a bridge expansion project. Even when it's illegal, punishments for heinous crimes often amount to wrist slaps, as when a multiyear killing spree by one of the largest wildlife poaching rings in US history earned most of its members fines of several thousand dollars. Enforcement is also scattershot and depends on government intervention. If someone starves his horse and the state declines to press charges, no other legal recourse exists. Citizens can't protect animals they don't own.

"Animals are treated as the property of their owners, rather than entities with their own legal rights," wrote the judge in a 1993 decision against Wise and several animal advocacy organizations that filed a lawsuit on behalf of a captive dolphin named

Kama. They alleged that Kama's transfer to the Navy for use in sonar research violated federal laws against selling marine mammals. Whether it did hinged on a spirit versus letter argument over the Marine Mammal Protection Act, but the argument never made it into court. The judge declared that Kama's human defenders were not themselves harmed by the dolphin's plight and thus had no legal standing to be there. As for whether Kama himself might have standing and could be represented by someone speaking for his interests, not unlike a young child—it was unthinkable.

Wise came to understand that fact as a fundamental obstacle to animal protection. "A thick and impenetrable legal wall has separated all human from all nonhuman animals," he wrote seven years later in a book titled *Rattling the Cage*. On one side are humans, legal *persons*, every one of whom is regarded by the law as having interests so important as to be called rights, with all the legal and moral power that entails. On the other side is every member of all other species on Earth. Their "most basic and fundamental interests—their pains, their lives, their freedoms—are intentionally ignored, often maliciously trampled, and routinely abused," declared Wise. And because they were not legal persons, their protections never rising to the level of rights, so it would remain until that wall was toppled.

Toppling that wall—or at least scaling it, obtaining legal recognition of at least one right for one animal, thereby transmuting them from thing to person in the law's eyes, and setting a precedent by which more rights could be recognized for more animals—became Wise's life work. With the Nonhuman Rights Project he developed a legal strategy to accomplish it. Late in 2018 the group filed a lawsuit on behalf of Happy, an elderly Asian elephant owned by the world-famous Bronx Zoo—and, argued Wise and his colleagues, a person, not only in the Kantian sense of the term but in a legal sense, with a right to be free.

◆　◆　◆

IN THE EARLY 1970S THE PLAINTIFF WAS CAPTURED from a herd in Thailand. No firsthand accounts survive of that day, but one can imagine the frantic wailing, the bodies crashing, the fear and rage, the grief. Elephant mothers and their children—calves, they are supposed to be called, a word that adds specificity but also emotional distance—love one another fiercely. Pregnancy lasts two years; young elephants nurse for several years, and they stay with their mothers for several years beyond that. Males leave their herds—their communities—as teenagers, while females stay their entire lives.

When Happy was kidnapped she was still nursing. She had spent her entire life within a few feet of her mother's gentle bulk, tended by the extended family who, if they were like other Asian elephant families, celebrated her birth with trumpeting. In the course of a few minutes all that was taken away. She and six other elephant children, already weighing more than an adult human but as sensitive and attentive as any toddler, were shipped to the United States; imagine yourself as a child, at the mercy of strangers of another species, in the cargo hold of an airplane. Their destination was a now-defunct wildlife park called Lion Country Safari, Inc. in Irvine, California. There they were named after the seven dwarves in *Snow White*.

Sleepy died within the year. The other elephants were sent to the Lion Country Safari in Loxahatchee, Florida, where they remained until 1977, when the corporation sold them to zoos and circuses across the United States. Happy and Grumpy, who had become fast friends and may have been siblings, ended up at the Bronx Zoo as stars of the Bengali Express Monorail, now known as the Wild Asia Monorail. They spent the 1980s being made to perform tricks and give rides. One of Happy's trainers called her "a more physical elephant than anything I've seen," a description suggestive of research on orphaned elephants. As adults they're prone to depression, unpredictability, and aggression, the hallmarks of what in humans is called post-traumatic stress disorder. Many scientists who study elephants now formally recognize the disorder in them as well.

In 2002 two other elephants, Patty and Maxine, attacked Grumpy and injured him so badly that he was euthanized afterward. The zoo separated them from Happy and put her with a younger female elephant named Sammie, who died four years later from kidney failure. Happy has been alone ever since, a creature who in the wild would wander tens of miles each day with her kin instead confined to a 1.15-acre plot of grass in warmer months and, in winter, to an indoor cage about twice as long as she is. Her only direct social contact is with her human caregivers. Precisely how these circumstances have affected Happy's physical and mental well-being is disputed, but it's safe to say that her life is very far from ideal. Anyone not invested in the institution of elephant captivity would call Happy's saga a tragedy.

But should it be considered a violation of a legal right to liberty, thus making Happy a legal person? That would be up to Justice Alison Tuitt, who would decide based not on the Constitution, nor on federal or state or local statutes, but on common law: that vast bulk of jurisprudence which covers everything else. Personal injury claims, for example, often fall under its auspices, as does privacy in the absence of regulations. Back when the US government refused to recognize same-sex marriages, some states had a common law tradition that did. Common law judges rely on precedents set by past decisions but are able to interpret them anew, and even to set new precedents, depending on changing knowledge and public values.

One of the most famous common law cases took place in England in 1772, a time when that nation's courts did not recognize Black people as persons. They were, as animals are now, considered property, deemed incapable of having rights of any sort. That changed after abolitionists demanded freedom for James Somerset, whom a customs officer named Charles Stewart had purchased from a Virginia plantation and taken back to England. In October 1771 Somerset escaped; several weeks later he was captured. Stewart locked him in a cell on a boat bound for Jamaica and ordered that he be sold into the early death of plantation labor.

Activists filed a writ of habeas corpus—a legal challenge to unlawful imprisonment—arguing that Somerset should not be considered a legal thing but instead a legal person with a right to be free. The Judge, Lord Mansfield, agreed. His decision laid the foundations for ending slavery in England. Wise wrote a book about *Somerset v. Stewart*, entitled *Though the Heavens May Fall*, referring to Mansfield's proclamation that justice should follow moral principle no matter what upheaval followed.

Wise drew upon the case as inspiration for his own work. On Happy's behalf the Nonhuman Rights Project filed a writ of habeas corpus, arguing that she be considered not property but a person, an entity with a single right: that of bodily liberty, thus freeing her from the arbitrary confinement in which she lived. The state of New York at least accepted this as a question worthy of serious consideration.

◆　◆　◆

THE FIRST HEARING, HELD IN A VAULTING COURTROOM in the Bronx, opened with lawyers from the Wildlife Conservation Society—which owns the Bronx Zoo—asking that Happy's case be dismissed because the Nonhuman Rights Project had already lost similar lawsuits involving captive chimpanzees. In each case the Nonhuman Rights Project had presented affidavits from primatologists who asserted that humanity's closest living relatives possess mental capacities that, in our own species, are the essence of autonomy, which in turn underlies our right to physical liberty. In each case judges deemed this fact irrelevant.

In the first, the judge declared that habeas corpus simply didn't apply to animals. An appeals court upheld that ruling. For someone to have legal rights, said the court, they must also be able to bear legal duties: a chimpanzee could not be expected to know the law. The judge in the second case was more sympathetic but declined to make what he called a "leap of faith," and again an appeals court upheld the decision, explaining that habeas corpus did not apply because the chimpanzee would be released to a sanctuary rather than set free. That would merely "change the conditions of confinement and not the confinement itself." The same could be said of Happy, for whom victory would mean life in a sanctuary instead of her original home or some wild place.

The third chimpanzee lawsuit was dismissed without a hearing, filed again, and heard by a sympathetic judge who quoted Supreme Court Justice Anthony Kennedy's observation—delivered in *Lawrence v. Texas*, the historic 2003 ruling that made anti-homosexuality laws unconstitutional—that "times can blind us to certain truths and later generations can see that laws once thought necessary and proper in fact serve only to oppress." In the meantime, though, she felt bound by the earlier rights-and-duties ruling.

Undaunted by these precedents, Wise, a slightly rumpled, affable man with gray hair and a steel-trap mind who seems born to be an underdog lawyer, proceeded to explain why they ought to be ignored. The conditions-of-confinement rationale ignored the essential question of whether an imprisonment was illegal or not. (Wise could also have noted that it technically sanctions the otherwise-illegal imprisonment of many elderly people. After all, freedom would only change their confinement's conditions from a jail to a nursing home.) As for rights-and-duties, that was a full-on disaster. If it held, Wise argued, it would strip rights from infants, young children, comatose adults, and people with mental disabilities who did not comprehend the law. "A tenth of the population of New York could not be persons," he said.

Wise then explained how the rights-and-duties ruling relied on a definition from *Black's Law Dictionary*—but when the Nonhuman Rights Project traced its origins to a 1927 essay by the legal scholar Sir John Salmond, they found that Salmond actually said rights *or* duties. Alerted to the error, the dictionary's editor corrected it. The updated version was published just a few weeks before the hearing. The Nonhuman Rights Project had, in a sense, already changed the very definition of person, and the original decision's foundations were crumbling.

Finally Wise came to the heart of the matter: who elephants are. The Nonhuman Rights Project's court filings included affidavits from five researchers—Lucy Bates, Karen McComb, and Richard Byrne from the University of Sussex; Cynthia Moss of the Amboseli Trust for Elephants; and Joyce Poole, cofounder of ElephantVoices—who together had spent more than a century studying elephants. Their names might not be familiar to the general public, but their presence was akin to an NBA all-star team joining an argument about basketball.

The affidavits summarized what is formally known about elephant intelligence, beginning with their brains. Like our own, those of elephants are unusually large relative to their bodies. We share with them a richly folded cerebral cortex, the brain's information-processing outer layer; large parietal and temporal lobes, which for us are integral to communication, perception, and comprehending actions; and a large cerebellum, so important for planning, empathy, and understanding others. Elephants also possess various specialized neurons that in humans contribute to attention, emotional learning, self-awareness, and self-control. "The physical similarities between human and elephant brains occur in areas that link directly to the capacities necessary for autonomy and self-awareness," testified the scientists.

Anatomy does not guarantee equivalence, but examples of elephant behavior followed, beginning with their performance on the mirror self-recognition test. As we considered earlier, the self-awareness of creatures who fail this test is too readily dismissed— but for those who pass, it's an empirical demonstration of a mental self-image. In an extraordinary coincidence, the experiment that established elephant self-awareness, announced in 2006 to great acclaim, was conducted by scientists visiting the Bronx Zoo. It starred Happy herself.

With a mental self-image comes the ability to set and pursue goals. With self-awareness also comes a richer sense of other beings' selfhood. The scientists' affidavits described elephants feeding individuals who could not use their own trunks to eat; a mother helping her infant climb a riverbank by digging her tusks into the mud to create steps; mothers pretending to be in estrous, otherwise known as being in heat, in order to show their daughters how to attract and respond to males.

The researchers also talked about elephant communication, a rich system of vocalizations and gestures they likened to human language. Their repertoire includes low-frequency rumbles that travel for miles, allowing them to communicate over long distances; at any given moment an elephant has a mental map containing the locations of their community's members. Experiments have also found that elephants can remember the voices of up to 100 other elephants, even when years have passed since hearing them. One wonders whether Happy recalls the voices from her childhood—and, if so, whether they've faded with time or remain as clear as on the day she was taken.

In controlled experiments, researchers measure elephant cooperation with tasks that require them to jointly pull a string to receive treats. In the wild, elephants cooperate to defend themselves against predators, including humans. They also make communal decisions about where to go and what to do. Their wandering is not aimless; they move between watering holes tens of miles apart and may travel more than 100 miles to places they've not visited in years, following routes memorized by elder matriarchs who serve as living libraries for their families. When an elephant dies, their kin may cover the body with vegetation and stand vigil against scavengers. They grieve.

"What we currently know is only a tiny fraction of what elephant brains are likely capable of," wrote the scientists, and those capacities form what in humans is cherished as autonomy. We choose the lives we want to live—within boundaries, inevitably, but those boundaries cannot be the confines of a cell. For an innocent person to be imprisoned is among the worst fates imaginable. And that, argued Wise, is what Happy's life amounts to: that of an innocent prisoner in something close to solitary confinement.

◆　◆　◆

HAPPY, AT LEAST, WOULD HAVE HER DAYS IN COURT. The second hearing commenced with a Wildlife Conservation Society attorney speaking for two industry groups—the Zoological Association of America and the Alliance of Marine Mammal Parks and Aquariums—who opposed the lawsuit. "This isn't just a case about Happy," he said, but "part of their campaign for numerous animals that is going to disrupt so many aspects of legal, social, and economic relationships within this state." The Nonhuman Rights Project is indeed open about wanting to free captive dolphins and whales, and one can envision activists seeking liberty for other species whom science provides with a similar evidentiary foundation.

"You are going to see a flood of litigation," said the attorney. "The courts aren't adequately equipped to deal with it. They're going to have arbitration lines drawn to say, yes for this animal and no for that animal. How are you going to do it? And then what's going to become of any animals that you free?" The Nonhuman Rights Project had arranged for Happy to live in a sanctuary, but other animals wouldn't be so lucky, he warned. Instead they would become burdens upon the state—and imagine if farmed animals were considered persons. The livestock industry could be shut down. People would go hungry.

These objections inevitably arise when discussing the Nonhuman Rights Project's cases, both in courtrooms and in the court of public opinion, and Wise acknowledged that claims to legal personhood may someday be made for other species. Courts would have to review the science and determine whether they deserved rights and, if so, which rights—but slippery slope arguments were no excuse for denying justice, and they were only there to talk about Happy.

Wise referenced Lord Mansfield and James Somerset, who "walked into court a common law thing, and walked out a common law person." He also invoked *Standing Bear v. Crook*, an 1879 decision that established the right of a Ponca man named Standing Bear—and thus all Indigenous people in America, who until then were not considered legal persons—to liberty.

"That was a human being," countered Tuitt.

Wise replied that species membership ought not matter. He offered a thought experiment: if a Neanderthal was discovered alive today, would that individual be a person? After all, Neanderthals are not *Homo sapiens*. Rather than clinging to a rationale that denies recognition to a being who clearly deserves it, look at who that being is. To do otherwise was to violate of the principle of equality. Like beings deserve like treatment: it's right there in the Fourteenth Amendment to the US Constitution, guaranteeing equal treatment under the law, and in Article I of New York's Constitution, and in countless laws and legal decisions and the fabric of social life in every modern democratic society.*

The third hearing resumed with deliberations about Happy's

---

* To invoke struggles for human equality in the context of animals may be problematic for some readers, especially when done by a white man. It has been argued that likening animal and human sufferings trivializes our own. It should also noted, though, that there is a long—and too little-appreciated—history of moral crusaders, from Gandhi to civil rights leaders Coretta Scott King and Angela Davis, who saw the plight of animals as existing on a spectrum with human oppression. At any rate, the comparison does not require the equivalence of animal and human injustice, but a consideration of shared principles.

present well-being. The Bronx Zoo explained that it followed standard industry guidelines and provided fine medical care. The guidelines, replied Wise, are woefully inadequate; though veterinarians tended to Happy's body, they could do nothing for her mind, which suffered from isolation and deprivation. He likened Happy to a patient in a prison hospital and presented an affidavit from Joyce Poole, who reviewed videos in which Happy repeatedly lifted her feet—suggesting either foot disease, which is a potential death sentence for an elephant, or mental distress.

The zoo's chief veterinarian had opined that even short moves within the zoo upset Happy, so transport to a distant sanctuary could be harmful. Poole claimed the difficulties of moves signified how traumatic captivity had been. The zoo said their vets were far better qualified than Poole, who knew Happy only from videos taken by riders aboard the Wild Asia Monorail.

That's indeed how I caught my only glimpse of Happy. Twice I visited the zoo, and each time it was not Happy on display in the 1.15-acre elephant enclosure but rather Patty, the zoo's other elephant. Like Happy, Patty had been captured in the early 1970s and now lived alone; in 1981 she endured the loss of Astor, her 17-month-old baby, to heart disease. (The *New York Times* covered the death. "He made it clear to zoo-goers that he didn't think they were very different from himself," said the zoo's then-director, William Calloway.)

The monorail cars are open on one side, with seats arranged to look in that direction. To see Happy, I had to hold my phone out my car's end and point it backward, giving a view of the low-slung buildings and dirt yards where Wild Asia's animals live while not in their display enclosures. On my first visit I didn't know this trick. The second trip yielded a video that I watched later. There, after the babirusas and before the Indian rhinoceros, as the monorail conductor pattered—"Patty is an Asian elephant and she is over 50 years old, but age is just a number to a lovely lady elephant"—was Happy, standing in the open doorway of a small windowless building. She faced inward, back to a tiny dirt yard, tail swishing.

At a glance the space resembled the scrapyards and industrial warehouses that ring the Bronx. It seemed like a miserable place for an elephant to live. But then—what could I know from a brief, indirect sighting? Perhaps I was projecting my own biases and narratives and expectations, yet it was also difficult to take the Bronx Zoo's public assessments at face value.

Would it be torture rather than salvation for Happy to leave her confines, or vice versa? It's hard to imagine how an enormous, extraordinarily intelligent creature evolved for companionship and movement could be mentally healthy in what amounts to a backyard with intermittent human contact. At the same time, diagnosis via video snippets could not be called conclusive. A reasonable solution, it seemed, would be to rule that an elephant is a legal person with a right to liberty, and then appoint some independent arbiter to decide what is best for her.

Again Wise stressed that Happy should be seen as a prisoner. She had been deprived of the ability to make choices, both physical and social. He spoke of the distances that wild elephants travel and their complex social lives. "These are elephants who, like you and I, engage in collective decision-making," Wise said.

"When they have to," Tuitt interjected. "If they are in a nice space, they don't necessarily have to do all those things."

It was unclear whether Tuitt actually thought that or wanted to prod Wise into putting an answer on record. Either way, the remark spoke to the gulf that exists between our understandings of humans and of other animals. Judge Tuitt would never have suggested that talking to friends and family is something one does out of utilitarian necessity, nor imply that living without direct human contact is perfectly tolerable so long as there's decent medical care and plenty to eat.

After a break for lunch, Wise tried a different approach. Imagine we didn't know Happy was an elephant and instead thought she was human, he said. Then the injustice of confinement would be self-evident. There would be no talk of welfare regulations or whether her captivity was comfortable. As the hearing wound down he once more stressed the importance of liberty, autonomy, and equality. "You have to follow those wherever they go," he said.

The Bronx Zoo's attorney invoked the precedents set in the earlier chimpanzee decisions. Should a judge disagree, he said, then it was a legislature's job to craft new laws. Granting personhood to Happy would upend the legal system. In short, though he didn't use the phrase, the heavens would fall.

◆   ◆   ◆

CRITICS OF THE NONHUMAN RIGHTS PROJECT AREN'T limited to opponents of animal rights. Some people share their values but feel that emphasizing a few select traits is unfair. "Some animals get in, but only because they are like us," writes the philosopher and legal theorist Martha Nussbaum. "The first door is open, but then it is slammed shut behind us." Great apes, elephants, and some cetaceans might get through, but other animals would remain "outside in the dark domain of mere thinghood."

Nussbaum did submit a brief in support of Happy, but her favored strategy is an adaptation of what's known as the capability approach, originally developed by Nobel laureate economist Amartya Sen as an alternative to metrics such as per capita gross domestic product. For humans it means prizing the ability of individuals to achieve what is important to them: for example, being healthy, having close friends, and fulfilling work. For animals it would predicate rights not upon cognitive similarities to humans but on what it means for an animal to flourish. Whatever a good life is for an elephant or a chickadee or a snapping turtle or a blue-spotted salamander should be the basis for their ethical consideration and ultimately rights. Of course one could argue that if courts hesitate to consider elephants as legal persons with a right to liberty, they'll be even less likely to grant a salamander's right to clean water—but that's a pragmatic quibble, not an objection to principle.

Still others think the focus on rights and justice is itself problematic. Not only might high-bar claims leave some species behind, but much of what's ethically important occurs in situations that don't lend themselves to adjudication according to universal principles expressed in law. This critique, associated with what came to be known as an "ethics of care," is rooted in academic debates that roiled the fields of feminism and moral philosophy in the late twentieth century. In the words of philosopher Karen Warren, care ethics make "a central place for values of care, love, friendship, trust, and appropriate reciprocity," and seek to understand those values as a basis for moral responsibilities expressed in the relationships we have. By this light, whether elephants or any other species should have rights is less important than what it means to be caring when their lives entangle with ours.

This approach also has critics. Once again, some animals might be excluded, and our obligations could be defined in self-serving ways. The differences between approaches based on rights or relationships also tend to be magnified by intellectual jockeying. They're not mutually exclusive. One might believe, for example, that dogs should have certain basic rights and that most human-dog relations—such as the dilemma of late-in-life euthanasia—are matters of personal judgement rather than law. As to the charge that grounding animal rights in similarities to humans would protect only a few exceptional species, Wise argues that this is a necessary first step. Once the precedent is set, conversations can expand to other species and what rights they should have. The rights of a sunfish would not be those of a chimpanzee.

It's certainly true, though, that legal personhood for animals will come slowly if at all and that ethics of care provide guidance for relations ill-suited to law. So one summer morning, seeking to root all this in something more tangible and fertile than the dry soil of academic discourse, I met with Kathy Pollard, a permaculture specialist and artist who lives in Orono, Maine, a few miles away from me.

We had struck up a friendship after I posted a photo on Instagram of a coyote I'd found dead beside a highway exit ramp near my home and carried to a resting place in the woods. Pollard thanked me; coyotes are dear to her heart, and as we exchanged mutual appreciations of those often reviled canines, she explained that several decades ago she was a student in the University of Maine at Orono's wildlife ecology program. The department also has close ties to the state's Department of Inland Fisheries and Wildlife, which at the time paid trappers to kill coyotes—something that was often done with neck-encircling snares in which coyotes might linger for days, with blood flowing into their heads but not back out, until finally blood vessels in their brains burst from the pressure. Learning that "made my soul sick," wrote Pollard as we corresponded.

There was no scientific basis for the killing, wrote Pollard; it was done to satisfy the hunting lobby, who claimed for themselves the deer eaten by coyotes. She left the program. The treatment of coyotes was a breaking point, but she had already been uncomfortable with an approach to wildlife management that treated animals as commodities—as beings lacking inherent rights over whom humanity could claim dominion.

"This of course collides with the traditional Indigenous beliefs which form the Original Instructions," the teachings "that lay out how to maintain reciprocal relationships with all who form the web of life," wrote Pollard, who is of Cherokee as well as German, Irish, and English heritage. "The term 'all my relations'"—an invocation widespread among Indigenous North American cultures, expressing a sense of kinship and also a moral responsibility to maintain respectful, balanced relationships—"includes humans *and* animals. When you think of all life as your relations the way you relate to all life is vastly different."

We met on a summer morning in a field behind the home of Lucy Quimby, the founder of the Bangor Land Trust. Pollard, who is in her early sixties and has a genial, slightly owlish air, wore denim overalls with soil patching the knees. She was already hard at work in the garden she planted there, not only to raise crops for herself but for food-insecure Wabanaki households. (The Wabanaki comprise five nations—the Maliseet, Mi'kmaq, Passamaquoddy, Penobscot, and Abenaki—whose homelands once formed much of present-day New England and Atlantic Canada.) Pollard is not Wabanaki, she clarified, but her daughter Ann is a Penobscot tribal citizen, and Pollard shares with them a cultural history of dispossession and genocide. The garden represented not only food but the reclamation of a stolen homeland and the lives it nourished.

As we walked through neatly tended rows of corn, bean, and squash, the traditional Three Sisters, still a few months from harvest, I mentioned how chipmunks kept eating shallot bulbs in my own garden. Pollard called my attention to dead pine boughs arranged over the squash; the sharp, unpalatable needles would discourage a neighborhood groundhog with a taste for young plants. There was warmth in her voice as she described him. Shooting or trapping was never an option.

In a larger planting, Pollard said, she would also grow milkweed and pollinator strips between the crops and celebrate butterflies who came to eat. She and a group of Native women were doing that at a farm along the Sandy River, in central Maine, where Indigenous people have recently returned to tending land their ancestors farmed for millennia. "In traditional Native ways, you would have a much larger field planted. And you would know that the periphery would be eaten by deer, raccoons, and whoever else might be hungry. Then the center"—protected by a tangle of prickly squash vines—"would be for you," Pollard said. "There's a sense that you're sharing, not excluding."

"We are taught from a very, very young age"—Pollard has a tendency to repeat words when she is impassioned, though her voice remained even—"not to hurt other life because it's inconvenient to us." That principle, she said, is contradicted by what she called a Judeo-Christian, go-forth-and-conquer mentality that also manifests in a sense of dominance over nature, of other lives as resources to be owned. Her gardening projects are a small refutation of that.

It's also possible, though, to live in environmentally light-footed ways without thinking of animals as fellow persons. I asked Pollard whether thinking of animals as fellow persons was necessarily integral to her spirit of giving. "Oh, absolutely," she said.

◆ ◆ ◆

EARLY ONE MORNING POLLARD TOOK ME TO PENJAJA-woc Marsh, a local wetland upon which development, in the form of commercial sprawl on one side and upscale housing on the other, has encroached. We walked there from the parking lot of a Walmart, crossing through late-summer meadows festooned with dew-jeweled spiderwebs and onto an abandoned railroad bed. Pollard planned to show me an area of beaver-tended willow she and her daughter Ann had found that spring. As if on cue, three beavers walking in single file crossed the path ahead of us, unexpectedly bulky yet quick, perhaps returning home after a night's activities.

I was reminded of an essay by Margaret Robinson, a Mi'kmaq sociologist who has written of her people's traditional relations to animals. The Mi'kmaq, who historically lived in the coastal regions of what is now Maritime Canada, just above Maine, had close cultural ties to the Penobscot. The stories Robinson recounts were perhaps inspired by the distant ancestors of these very beavers.

In one, "The Beaver Magicians and the Big Fish," a Mi'kmaq hunter follows snowshoe tracks to a lakeside wigwam. Inside he finds an elderly man and his family, who share a meal with the hunter and send him home with a gift of moose meat. When he unwraps it, he finds bark from a poplar tree; his hosts were a family of beaver, their wigwam a lodge. "Legends such as this one present human and animal life on a continuum," wrote Robinson. There were many such tales, with humans taking animal form or animals becoming human, conversing, sometimes marrying, and even raising families together. These were not necessarily understood as factual accounts but as parables containing a truth: that humans and animals share a common essence. Even the detail of the moose meat being turned into bark speaks to the mutual appreciation of a good meal.

The implications of that realization were nuanced and varied— among the Mi'kmaq and throughout the cultures that for much of human history shared similar beliefs—but broad themes can be found. For the Mi'kmaq, wrote Robinson, the phrase "all my relations" meant "that humans and animals both experience our lives in the first-person, overcoming fears, having adventures, falling in love, raising families, vanquishing enemies, and having a relationship with Kisu'lk, the Creator." One need not believe in Kisu'lk for the principle to hold.

We continued down the old railroad track. On our left was a shallow interstice between woodland and wetland; to the right, a vast sweep of cattails, the heart of the basin into which the surrounding watershed drained. The railbed itself was lined with trees and shrubs, and we passed several trees dotted with green herons who called to one another with almost catlike voices.

Pollard spoke of the marsh's importance to migrating waterfowl who stop there for rest and sustenance. The earliest gatherings likely occurred about 12,000 years ago, not long after glaciers receded from eastern Maine and the region's first human dwellers arrived. In the late 1990s, though, the marsh became the setting of a bitter development dispute.

The Walmart where Pollard met me was originally supposed to be built on the marsh's edge. The threat drove local conservationists into action, ultimately blocking construction and angering people who owned property nearby and reputedly hoped to sell it to developers at inflated prices. In 2002 some of them breached a 50-foot-long beaver dam, draining much of the marsh. Water remained, but far less than before. What was a shallow lake now has few stretches of open water. There's correspondingly less food to be had, and fewer birds stop during migration or make the marsh their summer home. In Penobscot, said Pollard, the word for wild rice, a favorite wetland staple of both humans and water-fowl, roughly translates to "the laughing berry"; when the voices of waterfowl had once filled the marsh, perhaps sounded like they were chuckling contentedly.

In 2003 adjacent property owners breached the dam again. This is illegal to do without a permit, but Pollard believes the deter-rence has continued. There's certainly incentive: environmental laws restrict building near a wetland's edge, so keeping water lev-els down increases the area available for development and thus a property's assessed value. And though tampering with dams is against the law, simply killing beavers is not. Throughout Maine they can be trapped without limit for half the year; the require-ment that trappers receive written permission from landowners before trapping on their property is even waived for beaver. Some-one dedicated to persecuting them would have few obstacles.

We walked out to where the marsh opened on both sides of the railbed. Red admiral butterflies and ruby-throated hummingbirds flitted around a stand of honeysuckle. Pollard backtracked and plunged down the bank, weaving through shrubs and into a thicket of low-growing willow just a little taller than we are. She invited me to look at the base of a tree: about two dozen stems, some thin and fresh green, others thick and mature, grew from central boles shorn above the mucky soil by a beaver's chisel-like incisions.

Many species of trees respond to pruning by growing in profu-sion rather than straight up, and for millennia cultures have used this to their advantage. The technique is known as coppicing; an area managed this way is a coppice. (A similar practice, with cuts are made around head height, is called—by happy coincidence—a pollard.) The result is not only a renewable harvest of poles but a sunlit bonanza of plant and animal life, with coppices often containing several times the number of plant and animal species found in comparable uncoppiced woodland.

"Willow basket-makers in England have vast fields that they coppice in the exact same way," Pollard said. That beavers coppice is known, though their work is only sometimes described as such in the scientific literature—perhaps because the term is fairly esoteric, perhaps because it implies an awareness and intentionality that beaver are not usually acknowledged as possessing. Their coppices are characterized as byproducts of feeding rather than an active form of management. Pollard didn't hesitate to use the word. "I've seen them coppicing," she said, "but never such a huge area."

Where we stood was only the edge of a 10-acre expanse of coppiced willows, all about 8 feet high, transected by canals where enough water remained for beavers to swim even though the region was in the midst of drought. Pollard and her daughter had found it strewn with freshly cut willow sticks from which beavers had peeled and eaten the bark. I asked if she thought the beavers understood what they were doing. "I do," she said. "I truly believe they know this is a perfect source of food, and that they keep it tended every year." She called it a form of permaculture.

Because these willows would replenish what the beavers harvested, they would not need to seek food on land, where predators could take advantage of their poor eyesight. So long as they

allowed shoots to regrow before harvesting again, they wouldn't need to abandon their homes when the right trees could no longer be found nearby. "They have ten acres of sustainable food here," Pollard said. "They can be here into perpetuity, essentially. I see it as akin to when hunter-gatherers became agrarian or added agriculture to their food system. People settled and were able to stay put." They are, she emphasized again, *aware* of this. The coppices are a strategy. "Generation after generation, they give it to the next," she said.

I thought of the three beavers who had briefly crossed our path and who I now understood to have been departing from the far end of the coppice, as well as the possibility that beavers in Penjajawoc were trapped in order to increase property values. I thought of colonial-era trapping that annihilated North American beaver populations—from around 400 million to a mere 100,000 at the turn of the twentieth century—and of how, until fur prices dropped a few years ago, approximately one-sixth of Maine's beavers were killed each year. All of it was something to see through a lens of personhood: not just in a legal sense but a moral sense, and encompassing not only obviously extraordinary creatures such as elephants but the humbler members of our shared world.

Thinking this way can be painful. It makes human cruelty and inconsideration, normalized by what Pollard called "our culture's paradigm of dominion over other living creatures," all the more piercing. To Pollard it's inspiration to do what she can to change the paradigm, one piece at a time: talking to homeowners near the marsh about pesticides and to the University of Maine's facilities managers about foxes, taking an ancient injured snapping turtle to a wildlife hospital, and urging the governmental committees and boards that guide development around Penjajawoc to find a middle ground that makes room for nonhumans. Each small action has a ripple effect, she believes, and eventually change will come.

Even if Penjajawoc's beavers are left alone, the pressure to develop the area will remain intense. But the Bangor Land Trust and other conservation organizations are trying to protect as much of it as they can. Lately they started collaborating with Pollard on what they call their Edible Landscape Project. She and Ann are planting a veritable cornucopia on the trust's lands: blackberry, raspberry, blueberry, and chokeberry bushes; plum, pear, and crabapple trees; wild grapes; and American chestnuts and black walnuts. When they've matured, and for centuries to come, anyone will be able to come and harvest food for themselves. For Pollard and the land trust, this is a way to help the region's food-insecure people and restore the rights of Indigenous people to live on what was, and still is, their homeland. But it's not only humans who will gather food, said Pollard. Anybody includes all our relations.

❖   ❖   ❖

TWO MONTHS AFTER HAPPY'S HEARING CONCLUDED, Judge Tuitt issued her ruling. "This Court agrees that Happy is more than just a legal thing, or property. She is an intelligent, autonomous being who should be treated with respect and dignity, and who may be entitled to liberty," she wrote. Furthermore Tuitt was "extremely sympathetic to Happy's plight," and found the arguments for moving her to a sanctuary to be "extremely persuasive." "Regrettably," Tuitt wrote, "the Court is bound by the legal precedent."

Not long afterward I visited Wise at his home in Coral Springs, Florida. We settled into chairs beside the shallow lake that abuts his neighborhood, keeping an eye on his dog, a furball terrier named Yogi. Moorhens dabbled in the shallows. Yogi, for his part, kept his eyes on two Muscovy ducks, black-and-white birds with fleshy red faces, historically native to South America and likely descended from ducks who were raised for food and escaped. They were using their own freedom to tease Yogi, approaching him and then waddling beyond the range of his leash.

Wise was philosophical about Tuitt's decision. There would

be more appeals, and he hadn't really expected to win. Defeats were inevitable when seeking such radical change. Not all losses were equal, though, and Tuitt's supportiveness was a triumph in itself. Perhaps the publicity—a Change.org petition had more than a million signatures; Congresswoman Alexandria Ocasio-Cortez and then-New York City mayor Bill DeBlasio had chimed in, too— would convince the Bronx Zoo to reconsider.

Effusive by nature, Wise bubbled with possibilities. More lawsuits might be filed in California, and possibly in Colorado and even Canada. The Nonhuman Rights Project was drafting model legislation for communities that wanted to recognize elephants and chimpanzees as legal persons within their jurisdiction.* The emergence of COVID-19 had put a temporary end to his breakneck schedule of teaching engagements and speaking appearances, but Wise and colleagues were still advising activists throughout western Europe and South America.

He was especially excited about developments outside the United States. A few months before Happy's first hearing, a lawsuit involving the mistreatment of cows led a court for the Indian states of Punjab and Haryana to declare that all animals are persons. The human residents of those states would legally be considered their guardians. "Animals should be healthy, comfortable, well nourished, safe, able to express innate behavior without pain, fear and distress," explained Justice Rajiv Sharma in his decision. "They are entitled to justice. The animals cannot be treated as objects or property." It was too soon to know what this meant in practice, but the winds of change were blowing.

---

* The Nonhuman Rights Project would go on to file habeas corpus petitions on behalf of three captive elephants at the Fresno Chaffee Zoo in Fresno, California, and five elephants at the Cheyenne Mountain Zoo in Colorado Springs, Colorado. As of this writing, both lawsuits are ongoing. In 2023 the group helped the city of Ojai, California pass a law recognizing the right of elephants to liberty within its borders. Shortly afterward, Wise, who was diagnosed with cancer in 2021, announced that he had stepped back from his duties to focus on his health.

Several months after that decision, Wise testified to the Constitutional Court of Colombia—that nation's equivalent of the US Supreme Court—in the case of a captive spectacled bear on whose behalf advocates actually won a habeas corpus decision. The court overturned that ruling, but a different fate awaited a habeas case in neighboring Ecuador, involving a woolly monkey named Estrellita. Although Estrellita died early on, the case wound its way through the legal system, ultimately reaching the Constitutional Court of Ecuador. They upheld a lower court's decision that Estrellita did not qualify for habeas corpus—but only because of the case's particulars and not because monkeys shouldn't have a legal right to liberty.

Monkeys indeed deserved that and many other rights. So did all animals. The resulting decision, which liberally cited a brief submitted by Wise and scholars from Harvard University's animal law program, was extraordinary. It put animal rights under the umbrella of Ecuador's rights of nature protections, which were written into the nation's constitution in 2008 to safeguard the rights of ecosystems to exist unimpaired by human activity. But whereas that law, like similar laws elsewhere in the world, considered animals strictly in terms of species and populations, the Constitutional Court decreed that it needed to consider rights for individual animals as well.

They had a right to exist; to not be hunted, captured, or trafficked; to suitable habitat or, for domestic animals, to adequate shelter; to freely express their natural behaviors and live free of fear or distress; and many more. These rights were not absolute, but they would have to be balanced with ecological imperatives as well as the rights of humans to meet their own needs. A harpy eagle who eats a monkey would not be a lawbreaker, and humans would still be able to farm animals. The court ordered the Ombudsman's Office to draft a bill of animal rights that would be presented to the national legislature. They would in turn be responsible for translating that bill into law—a process still ongoing as of this writing—and future jurisprudence would define and expand it.

As in India, it will take years, perhaps decades, to know what all this means in practice. Enforcement will be paramount: rights on paper are not the same as rights in reality. But even activists frustrated that the ruling didn't go far enough in curtailing the consumption of animals celebrated it as history changing. For animal rights to be enshrined in the constitution was both symbolically powerful and offered far stronger protection than mere welfare regulations. People would be able to sue on their behalf, and in considering whether an injustice had been committed courts would not look to regulatory minutiae but to the bedrock principles of rights. More rights might someday be secured.

That decision was announced shortly before the New York Court of Appeals, the state's highest court, heard the Nonhuman Rights Project's appeal of Happy's case. Five of seven judges upheld Tuitt's decision. They offered a stark contrast to the can-do spirit and moral vision seen in Ecuador. "Because the writ of habeas corpus is intended to protect the liberty right of *human beings* to be free of unlawful confinement, it has no applicability to Happy," wrote Judge Janet DiFiore in the majority's ruling. Happy was an elephant; she could not "assume legal duties and social responsibilities," which was a precondition for rights; and to rule otherwise "would have an enormous destabilizing impact on modern society," leaving courts to "face grave difficulty resolving the inevitable flood of petitions." Yet again: the heavens would fall.

◆　◆　◆

IT WAS DISAPPOINTING BUT EXPECTED. WHAT WAS UN-expected, though, were the two dissents, which were positively scathing. "The majority's argument—'this has never been done before'—is an argument against all progress," wrote Judge Rowan Wilson. Judge Jenny Rivera wrote that the rights-and-duties rationale had no legal basis and accused the majority of ignoring history and "preferring instead the comforting incoherence of its circular logic."

"History, logic, justice, and our humanity must lead us to recognize that if humans without full rights and responsibilities under the law may invoke the writ to challenge an unjust denial of freedom, so too may any other autonomous being, regardless of species," Rivera wrote. Though the majority prevailed, her words and Wilson's became part of the official record. Someday, perhaps, another judge would cite them in explaining why an animal did in fact deserve to be considered a person with a right to be free.

At one point during my visit to Florida, Wise showed me his office. One tall bookshelf was devoted entirely to titles on slavery, each of which he had read. Presently he was reading *Success Without Victory: Lost Legal Battles and the Long Road to Justice in America.* Written by Jules Lobel, president of the Center for Constitutional Rights, it's about struggles for justice—for abolition and the right of women to vote, against unjust wars and support for tyrannical foreign governments—that took decades to succeed. An underappreciated part of these struggles, argues Lobel, are so-called test cases. Often they lose; often they have no chance at all of winning. Nevertheless they inspire further struggle. They bend the long, slow arc of justice.

Wise had underlined passages throughout, but one chapter in particular, about antislavery litigators in the nineteenth century, was thick with notations. He paid particular attention to a section about Salmon P. Chase, a nineteeth-century American jurist. Chase disdained arguments based on procedural technicalities; he went straight to the matter's heart, characterizing slavery as immoral and unconstitutional. He earned the nickname of "Attorney General for Runaway Slaves," though he almost invariably lost. Judges would often say, as Judge Tuitt did of Wise, that though constrained by precedent they recognized the rightness of his arguments. Each appearance was an opportunity to change minds.

In the 1847 case of *Jones v. Van Zandt,* Chase took his attack on the Fugitive Slave Act all the way to the Supreme Court. He represented an Underground Railroad conductor found guilty of harboring fugitive slaves; defeat was certain, as the court had already found the Fugitive Slave Act, which allowed slave owners to recapture escaped slaves even in so-called free states, to be constitutional. The Chief Justice, Roger Taney, himself owned slaves. Chase was undaunted. Either the Declaration of Independence was a "fable," he proclaimed, or slavery was an abomination that no law could legitimize.

The case drew national attention; the *New York Tribune* published his argument, and Chase reprinted his legal briefs in a pamphlet distributed to every US Senator. In a few years Chase's arguments became foundational to the Republican Party that he helped found. Its first President, Abraham Lincoln, abolished slavery and appointed Chase to the Supreme Court.

"Chase's real plea in *Van Zandt*," wrote Lobel in a paragraph that Wise underlined heavily, with extra circles for emphasis, "was not to the court but to the public and history." I asked Wise whether losing ever became frustrating. Not at all, he said. He appreciates the slow, uneven pace of change. "I get aggravated when journalists just want to say, 'Well, did you win or lose?'" said Wise. "I'm thinking, no, we didn't actually lose. I mean, we lost the case—but you don't understand what we're trying to attain and how we're trying to attain it."

6

Citizen Animal

SEVERAL YEARS AGO I TRAVELED TO TORONTO, ON-tario, for a conservation biology conference. The city sits on the edge of Lake Ontario, and one evening I wandered down to Leslie Street Spit, a three-mile-long peninsula formed from construction debris dumped there during the last century. Upon that rubble grew an unlikely urban wilderness, which is now home to North America's largest colony of double-crested cormorants—large, fish-eating waterbirds with black plumage and bright blue eyes.

By historical standards Leslie Street Spit's 30,000 birds were not an exceptionally large colony, but they were still awesome to behold. Even from the distance I kept so as to avoid disturbing them, their physical presence was palpable. It recalled wildlife documentaries on wildebeest migration and emperor penguins, or the sci-fi film trope of a spaceship descending into an extra-terrestrial city, vast and teeming, at once alien but recognizable. Shorelines were black with their bodies; each tree might have several dozen nests, built from large sticks in a ramshackle fashion that belied their ability to withstand winds blustering off the lake. On the ground their nests were separated by only a neck's length. The landscape pulsed with the small movements of mates preening and tending to their young. Their low grunting voices made the air thrum.

Back in Maine, I had grown up disliking cormorants, regarding them as gluttonous competitors for the fish I wanted to catch. Eventually I stopped fishing, but well before that, I decided that their claim trumped mine. How could I begrudge them the fish they relied upon to survive but I caught only for fun? Ever since then I've felt an extra affection for them, a bit like when a former nemesis becomes a friend. I've also come to see my original antipathy as unwittingly reflecting a millennia-long vilification of the birds.

The Bible describes the Lord, while instructing Aaron and Moses on what to eat, calling cormorants detestable. In John Milton's *Paradise Lost*, Satan himself sits upon the Tree of Life disguised as a cormorant. Not every society felt this way; in China and Japan, other cormorant species have long been trained to catch fish, and among some Indigenous tribes of central North America, cormorants were one of the animals to whom people expressed a special kinship. Just as there was a Bear clan, a Sturgeon clan, and a Turtle clan, so was there a Cormorant clan. Settler society felt no such affiliation, though. They saw cormorants as destroying fish. In a culture where blackness "was an obvious and potent sign of difference and otherness," as the waterbird biologist Linda Wires wrote, the birds were regarded with "a special kind of hatred." Well into the twentieth century they were commonly called "n—r goose"—you know the word—and even if people who slaughtered cormorants were not necessarily motivated by racial enmity, the term speaks to the contempt many people reserved for them, and the low value placed on their lives.

Persecution of the birds reached its extreme in the late nineteenth and early twentieth centuries. Cormorants abandoned the breeding colonies—the ancestral communities—that were such easy targets for men with guns, who killed them by the boatload. Once common, they were reduced to a rarity; when in the 1920s a few cormorants were seen nesting in the Great Lakes, scientists disputed whether they had ever lived there before. The birds had been so thoroughly exterminated as to vanish from regional memory. There followed a brief rebound that lasted until the early 1940s, when reproductive disruption caused by the pesticide DDT decimated their populations again. Even where they persisted, they were a shadow of themselves. Far to the west, on San Martin Island off the coast of Baja California, where more than half a million cormorants nested at the turn of the twentieth century, just 5,000 remained by the late 1960s. A mere 125 pairs were estimated to survive in the Great Lakes. In 1972 the United States added cormorants to the species protected under the Migratory Bird Treaty Act and, with DDT off the market, their numbers started to rise again.

At sunset, as breezes stilled and Lake Ontario's expanse mirrored a long summer dusk, a sinuous black line appeared on the horizon. As it approached it resolved into approximately 70 cormorants who flew close to the water in a loose V-formation and veered upward at its edge, passing overhead in a rush of muscular, muffled wingbeats. No sooner had they faded when another squadron appeared, then another, and another. This was as moving a sight as the colony itself. These birds were returning home after a day's work—perhaps in the company of others from their neighborhood, I thought, with the family, friends, and acquaintances with whom they shared their lives. The procession had not abated when I departed a half-hour later.

What stories might these cormorants tell? Fishing stories, perhaps: they would know the waters for miles around better than any human angler. They would know where to find alewives—a staple of their diets—and also round gobies, rainbow smelt, and pumpkinseed sunfish, at different times of day and in different seasons. Could they also have some culturally-transmitted memory of those long, dark years when humans chased them from their homes and eggs deformed by pollution collapsed beneath the weight of incubating mothers? An unknowable and perhaps fanciful speculation, but cormorants can live for a quarter century, and only a few generations separated this metropolis from its founders. According to contemporaneous accounts there were six nests there in 1990, at the very beginning. Two birds per nest; in all, 12 *survivors*. What had it meant to them to find safety on this strange rubble outcropping, where despite the presence of so many humans they were actually left alone?

"Whose stories come to matter in the emergence of a place?" asked environmental philosopher Thom van Dooren and the late Deborah Bird Rose, an ethnographer, in an essay inviting readers to imagine how animals "understand and render meaningful the places they inhabit." It is good to conceive these stories, even if they cannot be investigated empirically; a beneficial form of anthropomorphism. How special the colony at Leslie Street Spit must have been to the cormorants who remembered a time before it. Perhaps they looked upon its society with something akin to pride. And what would it mean to include them—not just as individuals with rights, but with representation—in our own societies?

◆ ◆ ◆

NOT LONG BEFORE GOING TO TORONTO I ATTENDED A gathering of scholars and activists called Minding Animals. For someone with a background in science journalism, where the humanities merely supplemented stories about scientific discoveries and the researchers who made them, it was thrilling. There was an intellectual exuberance to the "animal turn," as this blossoming interest in animals is known, and a new term kept coming up at the conference: a "political turn," away from simply examining the ethics of our relations with animals and toward political organizing and including animals in institutional decision-making. Toward the idea, to put an American spin on things, that We the People ought to include animals as well.

That it doesn't can seem self-evident. How could politics *not* be a strictly human endeavor? Dogs might be like family, but they can't vote or write letters to the editor. Yet this attitude, argues the Dutch philosopher Eva Meijer, is rooted in a philosophical tradition of human exceptionalism. Aristotle defined humans as the only political animal, singularly capable of making moral decisions and articulating the reasons; subsequently animals were seen as voiceless, literally incapable of participating in the linguistic exchange of ideas at the heart of human politics.

In her doctoral thesis, titled *Interspecies Democracies*, Meijer excoriated this notion. "Other animals have languages and express themselves in many ways," she wrote, "but they cannot make themselves heard in the dominant political discourse because they do not speak in the language of power." She drew upon the scientific insights we explored earlier, about how a great many animals communicate in ways comparable to what we call language and live in complex communities where they negotiate— sometimes democratically—collective decisions. In light of this, argues Meijer, they should not only be seen as intelligent but recognized as a political group.

The Canadian philosophers Will Kymlicka and Sue Donaldson explore what this could mean in a book called *Zoopolis: A Political Theory of Animal Rights*, which was published in 2011 and, among people interested in animals and philosophy, was perhaps the decade's most influential work. At its heart is the proposition that—for animals as well as humans—moral status isn't enough. It's also important to be considered a member of society.

Prior to *Zoopolis*, Kymlicka had spent two decades theorizing about human social membership—specifically, about what it means to belong to multiple groups at once, especially in liberal, democratic societies. His *Multicultural Citizenship: A Liberal Theory of Minority Rights* and *Multicultural Odysseys: Navigating the New International Politics of Diversity* were not exactly airport-bookstore fare, but they were widely lauded within the world of political theory. It was Donaldson, to whom he is married, who introduced him to animal ethics when in the late 1980s a friend convinced her to go vegetarian. She in turn convinced Kymlicka. But it wasn't until the late 2000s that, at Donaldson's urging, the two considered the connection between animals and his ideas about human rights. Out of that came *Zoopolis*.

"Many people acknowledge that animals have an intrinsic moral status," Kymlicka explained at a Minding Animals panel, "Justice and the Political Status of Animals," so crowded that people ended up sitting on the floor. "But we've made much less progress on the idea that animals might rightly be seen as members of society with membership rights."

Membership rights, he explained, are those that come from being part of a shared society: my rights as an American citizen, for example, and then the rights that come with being a resident of the state of Maine, rather than the universal rights owed to me as a human being. It's the former sorts that figure most prominently in our lives, argues Kymlicka. Access to health care, the social safety net, necessary goods and services, political representation: all reside in group memberships rather than simply being human. And who qualifies for membership? Who gets to be part of what ancient Greeks called the *demos:* the recognized participants of what's now known as democratic society? In sixth century Athens, it was unenslaved adult men. Nowadays, at least in principle, it's every human being, but there's a tension at the heart of this expansion.

Some disability theorists critique how norms of citizenship implicitly presuppose a certain type of cognitive and linguistic sophistication. (Norman Rockwell's famous "Freedom of Speech" painting, of a man standing up and testifying at a town hall, always comes to mind when I think of this.) Children and people with cognitive disabilities are effectively excluded. This is unfair, say the theorists; all members of society should be considered citizens, and if it's presently difficult for them to participate in shaping their society and its laws, then we should figure out how to include them. This argument's implications for animals are clear.

Many of them are unquestionably members of our societies: most obviously domestic animals, whose very label speaks to their place in homes and workplaces, but also all those free-living beings whose lives are affected by our decisions and whose activities may affect our own. Animals, argue Kymlicka and Donaldson, deserve citizenship too. Of relevance is the way human citizenship is not a one-size-fits-all affair: for example, rights and privileges can vary depending on whether someone is native-born or an immigrant—and, within that, whether they're naturalized, refugees, or undocumented. Members of many nations and communities often share the same space: a busy street might contain a mix of natural and naturalized citizens, along with visitors from other countries and Indigenous people who belong to multiple nations at once. In this rich patchwork they find lessons for the status of animals.

In Kymlicka and Donaldson's model, domestic creatures—so intimately shaped by our wants and needs, and for whom we're uniquely responsible—would receive full co-citizenship. Their interests would merit special consideration; their situation would be akin that of young children or cognitively different people who might not be capable of conventional political participation but are represented by trusted people obligated to act on their behalf. Wild animals living in wilderness areas, who actively avoid humans and our settlements, would be seen as citizens of other nations. The wilderness—a concept critiqued by environmental historians as socially constructed and problematic in its elision of humans, but still a useful reference—would be their domain. Relations between us and wild communities would be governed according to established norms of cooperation, conflict mediation, and limited intervention. Once again humans would represent animals' interests but be guided by the principle that those wild creatures and their homes should be free from exploitation and invasion. Intervention might occur, but only for their own benefit, as with humanitarian aid.

Of course, members of wild communities often enter our own. Bird migration is the most obvious example, as with the solitary sandpipers who stopped to replenish themselves at the stormwater pond in Bethesda before flying on to Canada's boreal forest vastness. Kymlicka and Donaldson liken them to nomads, their sovereignty respected when crossing international boundaries, required only to respect the laws of their hosts. Once again humans would represent them and include their interests in institutional and political decision-making. And migrants are not the only free-living animals we encounter in our environs. There are those creatures, like pileated woodpeckers or coyotes, who survive in wilder pockets of settled areas; all those animals who thrive in urbanized ecosystems, such as white-tailed deer, house sparrows, and pigeons; and also the so-called ferals, the escapees and descendants of animals bred by humans.

These billions of liminal creatures are especially interesting to Kymlicka and Donaldson. They're "affected every time we chop down a tree, divert a waterway, build a road or housing development, or erect a tower," they write in *Zoopolis*, yet "from a legal and moral perspective, they are amongst the least recognized or protected animals." Conservationists and animal advocates alike often overlook them, perhaps in part because they don't fit into a tidy conception of cities as human spaces and wild creatures as not truly belonging where we reside. In Kymlicka and Donaldson's model, they are not full citizens, which entails a level of human interaction they're not seeking, nor members of sovereign nations, a designation that fails to reflect how entangled they are with our own lives. Instead, these beings—the starlings and cottontail rabbits of the stormwater pond, the cormorants on Leslie Street Spit—would be considered denizens, not unlike immigrants who are not citizens but are still accepted as members of society. They don't qualify for citizenship's full benefits but merit consideration and some degree of representation in society's deliberations.

All of this might sound impractically radical as well as academic—something for ivory-tower theorists oblivious to the profound disagreements that exist about representation for fellow humans, much less animals. Yet it's still vital to grapple with what it means to see animals as fellow society members.

It's now widely accepted in environmental and conservation circles that humans and nature should not be understood as mutually exclusive, intrinsically opposed categories, with human presence inevitably blighting a nature that flourishes only in our absence. Contemporary nature writing is animated by a spirit of imagining new relationships of mutual thriving—yet animals remain largely invisible in politics and governance, even theoretically. When in her best-selling *Braiding Sweetgrass* Robin Wall Kimmerer writes of "the democracy of species," the possibilities of that beautiful phrase remain unexplored. In this way the human–nature divide remains intact.

Though it's presently difficult to envision for some domestic animals—not pets, perhaps, but certainly those used for food, labor, and research, which are industries that animal citizenship would challenge—it wouldn't be hard to include wild animals in the way Kymlicka and Donaldson envision. Killing them for sport or convenience is unjust; so are the harms of pollution, vehicle collisions, and climate change. Respecting and representing their interests isn't so complicated, and I suspect many nature- and animal-loving people would like to see it. The practice just needs to catch up with the theory—and, far from being radical, these ideals are already, at least in some places, being realized.

◆ ◆ ◆

IN THE MID-2000S SOME TORONTO RESIDENTS WANTED the cormorants gone. They blamed them for killing the trees where they roosted with their acidic, nutrient-rich guano. This was unquestionably true, albeit at a scale dwarfed by the construction of new subdivisions and shopping centers in the region. But people were particularly fond of Leslie Street Spit, which is home to the popular Tommy Thompson Park and its 11 miles of urban wilderness trails, and it was far easier to kill cormorants than challenge property developers.

Cormorants, after all, don't typically inspire the sympathy that blue herons or bald eagles do; they're not so beautiful or majestic that it's easy to care for them without really knowing them. To get a better sense of who they are I asked Gail Fraser, a biologist at York University who has spent years studying the colony, about their daily lives.

What stands out to her, she said, is their devotion as parents. Like many bird species their babies are born naked and undeveloped, requiring constant attention, but this dedication lasts even longer than she expected. At the end of summer it's not uncommon to see parents remaining at their nests, waiting for visits from otherwise-independent chicks. "The adult is waiting there. And waiting there and waiting there," Fraser said. "And then the chick flies in," often clumsily, banging into branches, "because they're still learning to navigate and fly. But they arrive, and they immediately start begging, and they hardly have to beg at all." Mom or dad is ready with a meal.

Yet even Fraser's knowledge is not intimate. I asked her whether cormorants used the same nest each year: she thinks so, although she's not sure. To reduce disturbance, she's put the colored leg bands that biologists use for identification on only a few birds. That seems emblematic of how rarely we know these birds as individuals, much less at a personal level.

I found one person who knew a cormorant well, though: an ecologist named James Ludwig, the cover of whose book, *The Dismal State of the Great Lakes,* is graced by a cormorant he found in the summer of 1998 while researching the long-term effects of pollutants on waterbirds. He was on Naubinway Island, at the northern tip of Lake Michigan, finishing yet another depressing field season of cataloging birth defects and reproductive disorders. Within the eggs he sampled, one-third contained deformed embryos. When he found a fledgling who clearly would not survive, he would euthanize the bird for further analysis.

He had already killed 28 cormorant and Caspian tern chicks—"death had been in my hand, and by my hand, daily for two months," he later recalled—when he found a baby cormorant whose upper bill curled into a near-complete circle, like a cartoon boar's tusk. Ludwig had never seen such a spectacularly disfigured bird. "Something about her really struck a chord," he told me. He knew it was his duty to kill her, and chided himself for being so unscientifically soft-hearted, but he looked into her eyes and decided to keep her alive instead. He named her Cosmos.

At the time he thought the gap between their species precluded any sort of relationship, and theirs began—appropriately for a baby—with incessant hunger, demands for attention, and prodigious amounts of poop. Cosmos grew fast, and soon Ludwig started taking her along when he testified to legislatures or spoke in public about pollution in the Great Lakes. One evening, as they drove home from a meeting in Chicago where his request for research funds had been denied, Cosmos hopped onto a despondent Ludwig's shoulder and started to nibble his ear. It wasn't so different from what a pet dog or cat does when sensing their human's distress, or from the way cormorants preen their partners. Instead of feathers, Cosmos tended to Ludwig's sideburns.

That became a drive-time ritual, and the memory of their long trips across the upper Midwest remains precious to him. Over the course of nine months Cosmos made, by Ludwig's count, 27 television appearances, including a visit by a news team from Japan, where she graced the cover of a middle-school textbook; joined him in a dozen classrooms and 29 meetings of local environmental groups; and accompanied him to the state buildings of Michigan and Wisconsin, and also to the US House of Representatives. She posed for photographs and seemed to delight in attention, as befitted such a social species. Whenever Ludwig was especially anxious or unhappy, she would invariably preen his hair and nibble behind his earlobes. Cosmos herself liked to have her feet stroked and her neck rubbed. Sometimes she fell asleep on his shoulder.

At home she would join Ludwig on the backyard bench where he liked to read and turn the pages of scientific journals with her beak. She could be possessive, too, becoming upset if someone else tried to sit with them. "I think she looked at me as her parent," he said. Often she went for a dip in his pool; at those moments, said Ludwig, she was transformed. Never mind that she was a sickly bird upon whom the legacy of dioxin dumping was written not only in genes that regulated beak development, but in a suppressed immune system that left her chronically vulnerable to infections that Ludwig and his wife Kay could barely control with repeated cycles of antibiotics. In the water she became a glistening bullet, streaking from one end to the other faster than Ludwig could run.

But that chemical legacy would not be denied. When the next summer arrived and Ludwig prepared for another field season, Cosmos was recuperating from a chest infection and foot sores. Soon after he left she developed a recurring eye infection. By midsummer she was weaker than ever. During a break in his work Ludwig returned and she seemed to regain some of her strength, but her health declined again when he returned to the field. She stopped eating altogether. One early August afternoon Kay found Cosmos beside the pool, in her favorite sunning spot, her lifeless bill grazing the water's surface.

Ludwig has an Upper-Midwestern voice, amiable but also reserved, and grew up well before it was widely acceptable for men to expose their emotional vulnerability. Still, when he tells me this, I can hear the tremor in his voice. "I felt the same way that I felt when I lost my 15-and-a-half year-old border terrier," he said. "Like an essential part of my life was gone. And it took a long time for me to really deal with it."

Because Ludwig was away when Cosmos died, Kay put her body in a freezer, and when he returned he left her there. Rather than confront his own feelings he sunk into work; he tried to convince himself that Cosmos was just a bird.

It took him two years to decide what to do with his friend. He and Kay finally took her body to High Island, to a promontory they consider to be the most beautiful spot in Lake Michigan. There, as the region's Stone Age inhabitants once did, they faced her body to the east to catch the rising sun.

Below her was a colony of gulls and terns, and as Cosmos never had a nest of her own, they built her a nest of dune grasses. In place of eggs, they put stones in it. From nearby wildflowers he gathered roots and seeds: Cosmos had died before her own colors came in, but these would bring colors each spring. Over her body they read a classic passage from the naturalist Henry Beston's *The Outermost House*:

> *We need another and a wiser and perhaps a more mystical concept of animals. . . . We patronize them for their incompleteness, for their tragic fate for having taken form so far below ourselves. And therein do we err. For the animal shall not be measured by man. In a world older and more complete than ours, they move finished and complete, gifted with the extension of the senses we have lost or never attained, living by voices we shall never hear. They are not brethren, they are not underlings: they are other nations, caught with ourselves in the net of life and time, fellow prisoners of the splendour and travail of the earth.*

Cosmos was special—and yet, in a sense, no more so than any other cormorant. Circumstances brought her and Ludwig together, but had Cosmos been born healthy and never caught his eye, she would have been just as valuable a being. And even though I can't tell the cormorants of Leslie Street Spit apart from one another, and the biologists who study them struggle to identify individuals, much less know their personal histories, each of those birds—those denizens, in the language of Kymlicka and Donaldson—is special, too.

Bear that in mind when considering the calls for their removal—which is to say, their killing. As these intensified, up stepped the Animal Alliance of Canada, an animal protection group cofounded in 1990 by Liz White, a former nurse who had previously worked for the Humane Society. In 2005 she also helped found a political party now known as the Animal Protection Party of Canada. White and company didn't actually expect to win elections, or even come close, but by running candidates they might help shape public conversation. They were "there to defend the rights of animals and make people understand that the planet is a finite entity, and that we're all in it together," White told me when I visited their headquarters, which double as a shelter for rescued cats and dogs. Adorning the walls were photos of animals they'd helped: seals, cormorants, cows and sheep, a turtle kept in a bucket for 20 years for whom they'd found a sanctuary home.

White, who still exudes a nurse's brisk competency, offered me a slice of vegan cake and explained how the Toronto and Region Conservation Authority had been ready to start shooting cormorants at Leslie Street Spit. This was predictable; even in an age of relative protection, about 40,000 cormorants were being killed with government permission every year in the United States, and some 32,000 had been killed in the Great Lakes region of Ontario. In the United States, this was presented as necessary to protect fisheries—a justification that, nearly a decade later, would fall apart when challenged in court—while in Ontario cormorants were killed mostly in the name of protecting vegetation around their roosts.

Local conservation groups might not have liked the proposal, but it wasn't as though the species was in danger. Instead it was the animal protection community that made a stink. After Animal Alliance and their allies threatened a lawsuit and protests, the TRCA agreed to create a management group that included both cormorant foes and advocates. In other words, the Leslie Street Spit cormorants now had representation. Ultimately the parties compromised, setting aside three areas of the spit where cormorants are welcome. Outside them the park's managers remove newly constructed nests in winter, when birds have already flown south, and in springtime at the start of nesting season, although they'll refrain if eggs or chicks are present. Within the cormorant-friendly zones they've encouraged roosting on the ground rather than in trees, at first by putting cormorant decoys in artificially constructed shoreline nests, setting out nesting material, arranging fallen trees to provide protection, and playing cormorant vocalizations over loudspeakers. Over the next eight years, the number of tree-roosting cormorants fell by nearly half while the population of ground nesters rose ninefold.

"It's the only progressive management approach to cormorants that I've seen," said White. "If you decide to put your mind to trying to solve a problem that isn't the automatic 'get out your gun and shoot the birds,' you can come up with some pretty creative stuff." In 2018, Gail Fraser, the biologist, and colleagues gave the plan their scientific seal of approval, writing in the journal *Waterbirds* that it "allowed for the sustained existence of a thriving colony."

This may not seem remarkable: it's what advocacy groups do, after all. Yet it shouldn't be taken for granted. People spoke on the cormorants' behalf in dialogs that—beyond the act of voting—are the foundation of political life, and then represented them in decision-making processes. It's a powerful, subtly profound example that shows why thinking of animals as citizens isn't so radical: it's an outgrowth of things we already do, but usually without framing them in that language. And the possibility exists to accomplish so much more.

◆ ◆ ◆

REPRESENTATION FOR LESLIE STREET SPIT'S CORMO-rants came together in ad hoc fashion, but the principle could be applied more formally. And although there's more to politics than voting, it remains the essence of democratic representation.

While the Animal Protection Party of Canada has never had a candidate elected—White's political high-water mark, set in a 2008 House of Commons election, was 0.52 percent of the vote—in Western Europe, animal advocacy parties have had more success. In 2021 the Dutch Party for the Animals received 3.8 percent of the general-election vote, giving them nine seats in the country's national legislatures and one of the country's 26 seats on the European Parliament. In Portugal, the People-Animals-Nature party won a European Parliament seat in 2019 and four national Assembly seats. There are now more than a dozen such parties worldwide.

From my vantage in a country dominated by two political parties where the future of liberal democracy seems uncertain, these developments read almost as dispatches from another planet—yet there are ways of working within this system, too. One model is offered by Voters for Animal Rights, a group in New York City that organizes people who care about animal issues into a voting bloc that can influence policy and help elect sympathetic candidates. VFAR's legislative victories include citywide bans on the use of wild animals in circuses and on the sale of foie gras, which is produced by force feeding ducks and geese; and the imposition of tougher penalties for capturing wild birds, an initiative prompted by the netting of city pigeons for use as live targets at gun clubs.

Their most significant accomplishment, though, was the push to create a position dedicated to animal welfare within the mayor's office. The role started in 2015 as a liaison within the city's Community Affairs Unit, one of many tasked with connecting City Hall to especially important or vulnerable groups. (There were also liaisons to New York's Jewish, Muslim, and LGBTQ communities.)

In 2019 it became permanent, safe from the vagaries of discretionary budgets and new administrations. It is the first position of its kind in the United States, perhaps anywhere: a city official whose job it is to care for the well-being of animals and represent their interests within government.

"I am here to advocate internally for my constituency—which is the animals themselves, although the animals don't vote," said Christine Kim, who under former mayor Bill DiBlasio headed what is now known as the Mayor's Office of Animal Welfare. The OAW is not an agency unto itself. Instead it collaborates with other agencies when their duties overlap with animals: with the Parks Department, for example, on a deer sterilization program, or with the Department of Health and Human Services to construct new animal shelters.

The OAW is as small as an office can be. Kim was its only employee. Nevertheless, she and Rachel Atcheson, who preceded her on the job, did quite a bit. In addition to the accomplishments already mentioned, they promoted Meatless Mondays; pushed for an end to school egg-hatching projects; switched to nonlethal management of Canada geese in city parks; expanded the city's network of animal shelters, one of which will house a wildlife rehabilitation center; and, during the COVID-19 pandemic, set up a service to help people care for their pets while stuck in quarantine or hospitalized. The OAW also consults with other officials and departments, and animal advocates get in touch, too. "It gives us a direct line into city government," Allie Feldman Taylor, VFAR's founder, told me, and it's a direct line for citizens as well. Were someone concerned about something as basic as the trapping of turtles in their local park, Kim said they could call OAW.

The OAW role represents only a glimmer of what's possible. Janneke Vink, a law professor at Open University in the Netherlands, has suggested other political roles: animal ombudsmen, political advisory boards for animal issues, constitutional requirements that legislators consider nonhuman interests, and state-funded animal rights lawyers. There might be animal representatives on community boards, city planning committees, and even city councils. Muhamed Sacirbey, a former United Nations ambassador and International Court of Justice agent, thinks the U.N. should have an animal ambassador.

There are institutional possibilities, too. Laura Bridgeman, founder of cetacean advocacy group Sonar, wants representatives of whales and dolphins appointed to the International Whaling Commission and included as stakeholders in Marine Protected Areas, alongside the human communities whose interests are represented. The legal scholar Karen Bradshaw envisions how land can be designated for co-ownership by animals and managed by trustees with a fiduciary duty to represent their interests. Every large institution—an arboretum, say, a university, or a company—might include an animal advocate on its board of trustees.

Examples of political representation for animals can also be found in Indigenous traditions. Leanne Simpson writes of Ojibway First Nation traditions in which "animal clans were highly respected and were seen as self-determining, political 'nations' (at least in an Indigenous sense) to whom the Nishnaabeg had negotiated, ritualized, formal relations that required maintenance through an ongoing relationship." There were treaties with animals—best understood not in terms of formal legal agreements, but rather sets of mutual relations and responsibilities. Ojibway communities gathered to discuss issues involving their clan animals, with clan members representing the interests of the animals in question. Their connections ran deep. "All animal nations, fish, and birds are represented in the clan structure," said Aimeé Craft, an expert in Anishinaabe and Canadian Aboriginal law at the University of Ottawa. "We were told that we were descended from those beings. As their relatives, we represent them. Even if we're not directly inviting the fish to the table, there are fish relatives that are part of that governance."

All the examples Craft mentioned involved animals upon

whom Indigenous people had relied for sustenance and where responsibilities and reciprocity pointed toward actions that would nourish habitat and populations across time. One could argue that this type of representation is already embodied by state-level wildlife agencies, but there are important differences. Wildlife management decisions tend to be focused on particular species: when making decisions about deer, coyote interests are not represented, and coyote advocates are usually excluded. Wild animals are understood to be property of the state, not their own nations. Serious talk of kinship is not part of the management equation. And though extrapolating these principles to some contemporary contexts, such as cities or suburbs, or to species upon whom humans don't rely, might seem challenging—"I'm trying to figure it out for myself," said Craft, who believes that representing animal interests is the right thing to do. "One of the things I've been advocating for is having a seat at the table for them," she said.

It's fun to imagine how a multispecies congress might be run: my stormwater pond's management committee could have representatives for chimney swifts and swallows; robins and frogs; and rats, raccoons, and rabbits. The frog people might be especially sensitive to issues of pesticide use in the neighborhood, as would swift and swallow delegates. The swifts would certainly want to know about chimney maintenance schedules. Rabbits would have much to say about mowing. Robins could accept the necessity of removing nonnative English ivy vines before they overgrew trees but would ask that they be cut gradually so as not to starve the creatures who eat their fat-rich berries all winter. A bit fanciful, maybe, but the point of invoking these traditions is to expand the imaginative space and to remind us that giving animals a political voice is hardly unprecedented. There are histories in which this was absolutely normal. From another perspective, *not* having animals represented is the aberration.

The policies pursued in New York City also embody but a fraction of what could be done. Countless animal- and conservation-related innovations and best practices have been developed; they simply need to become norms rather than scattered bright spots. One of my favorite examples comes from Gail Fraser, who when not studying cormorants participates in Ontario SwiftWatch, a program organized by the conservation group Birds Canada in which bird-loving citizens document chimneys used by swifts to roost. Their reports are tracked by Toronto's City Planning Division, and when a property owner applies to develop one of these locations, the application is flagged. The city works with them to avoid construction during roosting season and, even better, to keep the chimney intact or build a new one. Something similar happens in London, England, where licensed bat surveyors review new construction and renovations for the presence of bats, each species of which is protected in Europe. The surveyors recommend how to avoid disturbance and make each building as good a home as before, or even better.

Chimney swifts and bats are the exception, though. Animals who are not endangered or threatened tend to receive little attention in planning and development. What if this were done everywhere—and not just for protected species, but common creatures, too? At present, the latter often don't appear in environmental impact statements, except to say that impacts upon them are not especially important.

If that possibility raises the dreaded specter of more regulation and red tape, the practice need be complicated. When I joined the local commission that helps regulate development around Penjajawoc Marsh in Maine, we pushed for a proposed solar farm to include wildlife-friendly fencing, mowing regimes, and construction methods. In doing so we drew on an emerging literature of animal-friendly design that also includes bird-friendly windows, animal-safe lighting, wildlife road crossings, and attention to creating corridors between nature-rich areas. Lesley Fox, executive director of animal advocacy group the Fur Bearers, pointed me toward Oakville, Ontario, where conservationists and animal advocates together developed the city's wildlife and biodiversity strategies. They include everything from road ecology and meadow management to discussion of specific beaver colonies and plans for reducing conflict with aggressive mother geese by erecting temporary fencing beside their nests.

There are, in short, many ways to further the interests of animal citizens, denizens, and members of wild nations whenever society makes decisions that involve them. This isn't to say that representation is a panacea: it could conceivably be turned into a self-serving process. Institutional Animal Care and Use Committees, which ostensibly represent the interests of research animals at universities but often rubber-stamp inhumane or unnecessary proposals, are a cautionary tale. Yet the same caveats apply in human contexts, too. A child's guardian might be selfish, but mechanisms exist for ensuring that the guardian acts in good faith. And though tensions could exist between animals—might bear and salmon have different views on salmon management? What happens when habitat development or even climate change harms some animals but benefits others? Should the interests of animals who might thrive in a newly balmy Arctic be voiced?— that's still no reason give up.

"All of those voices would have to be weighed," said Aimeé Craft when I asked her about conflicting animal interests, with "the understanding that to the best of each party's ability, you would need to look after the collective well-being." The same goes for situations in which human and animal interests collide. Adjudicating between many interests and values isn't always easy, but it's what democratic institutions were designed to do. We try to consider more viewpoints, encourage more deliberation, and to be fair. Having *less* representation is never the answer.

◆　◆　◆

SOMETIMES PEOPLE WONDER IF IT'S TRULY POSSIBLE to know what's best for animals or what animals want. To me this misgiving is overblown. Certainly it's important to scrutinize claims made on animals' behalf, and there are difficult cases: given the choice, would a female bison in Yellowstone prefer to have her herd's size regulated by contraception or by shooting, with a chance at long life coming at the expense of reproduction and vice versa? Usually, though, it's straightforward. It's best for a warbler to not ingest pesticides. A turtle would rather not be run over by a car. Nobody wants to lose their home.

It's not difficult to imagine something fairer and more representative than what is happening now to cormorants. Late in 2020, the US Fish and Wildlife Service announced that it would allow up to 121,504—roughly 13 percent of their population—to be killed each year, while also making it easier to get the necessary permits. What's more, it would now be permissible to kill them merely for having an impact on wild fish populations, or even being perceived to. "The mere presence of double-crested cormorants consuming fish is the conflict," wrote Linda Wires, the waterbird biologist, in a comment submitted on the plan, which she said would "legitimize irrational attitudes and authorize destruction of birds based on myths and misperceptions."

"There have been hundreds of studies done looking at cormorant diets and trying to determine what the impact on fish populations is," she told me, "and it's not very easy to figure that out." It might *seem* straightforward—they eat small fish!—and yet it's hard to know whether their predation actually drives populations down, or in fact creates ecological openings for other fish, so that population-level impacts are negligible. Most studies, said Wires, show that when prey fish decline, it's for other reasons. In the Great Lakes, these include pollution, changes in nutrient levels, and human fishing pressures. In the western United States, dams have blocked salmon from the streams and rivers where they spawn. "But it's so convenient," said James Ludwig. "Blame the black bird. You've got to show your constituency that you're doing something for them."

That doesn't mean cormorants don't ever cause fish populations to decline; sometimes they do. But might cormorants enhance a system's productivity in other ways, helping other species thrive as a consequence? One fishing trick is to cast beneath their colonies, in waters where their nutrient-rich feces nourish invertebrates consumed by other fish. When we view cormorants—and by extension the lands and waters where they live—primarily in terms of fish production for humans, these complexities are excluded. And when the birds' impacts are more pronounced, does that mean they should die for it? What does it say when a species is allowed to live only in the margins of a nature that we've claimed for ourselves?

Among the scores of people involved in regional meetings leading up to the final USFWS decision there was not a single representative of a conservation or animal advocacy group. Nor was there an ethicist who might speak to issues that, as environmental ethicists Chelsea Batavia and Michael Paul Nelson wrote in an article in the journal *Waterbirds*, had been so spectacularly missing from cormorant management. "If cormorants were regarded and valued not merely as means, but also as ends in themselves, their interests could not, in good conscience, be so completely overlooked," they wrote. The species' survival was not yet at stake—but the more pressing issue, they said, is whether "in killing cormorants to eradicate competition (or perceived competition) with humans, we are appropriately sharing common resources." Nobody had asked what it means to share.

Yet even that was arguably fairer than what happened in Canada, where in 2018 the Ontario government proposed an essentially unlimited cormorant-shooting season: up to 50 per day per hunter from the beginning of nesting season until long after they migrated. It was simply a plan for slaughter, so egregious that it provoked an outcry not only from animal advocates but from conservationists and government scientists. After a year's delay the plan was reconfigured. There wouldn't be killing during the nesting season, but it would proceed apace beginning in mid-September. The daily limit was reduced to 15 but, unlike other hunted birds, kills didn't need to be reported and could be dumped in a landfill. It was still a plan for extermination.

The darkness may be coming for these black birds again—but Liz White assured me that Animal Alliance and others who have protested will continue to do so. And on Leslie Street Spit, the last great cormorant community in North America, they will remain safe. There people speak for them.

III

CLOSE TO HOME

7

Redeeming the Pest

EVERY SO OFTEN ONE MEETS A CHILD WHO IS EXCEP-
tionally kind to animals. Perhaps you were one. Brad Gates was;
whenever someone in his suburban Toronto neighborhood of
Scarborough found an orphaned fledgling, Gates was the kid who
would take care of the bird. He raised rabbits, squirrels, and pigeons, and he spent long idyllic summer days exploring the ravines
that ran through the town in fingers of topographically uncon-
querable wildness.

In April of Gates' senior year in high school, he saw a newspaper ad from a local wildlife removal company that was giving away
baby raccoons. It was 1979, before owning wild animals became
illegal, and soon Gates was the proud foster parent of a three-week-
old ball of fur and need whom he named Mandy. He spent the last
summer of his adolescence raising Mandy to hers. She became his
constant companion, accompanying him on his ravine wander-
ings and to soccer games in the park; he did his best to introduce
her to the world in which she'd soon need to survive. At summer's

end, with university approaching, he started leaving the door to her backyard cage open at night. By the time school began, she had stopped coming back.

One evening about two years later, as Gates pulled into his driveway after a day of classes, a raccoon dashed across his headlights. Though he had not seen Mandy since letting her go, he still stepped out and whistled the tune he'd once used when it was feeding time. The raccoon was Mandy. She dashed across the lawn, ran up his leg, and perched on his shoulder, and together they raided his refrigerator and had dinner on the lawn. It was the last time he saw her.

Gates now remembers that as a defining moment in his life. He decided that he wanted to work with wildlife, especially raccoons, and got a job with a wildlife removal company. It was not what he expected, though. Raccoons were treated as vermin. When someone reported an animal in their attic, the company pumped formaldehyde through a fogging machine to drive them out. The technique often didn't work as intended, and the poor creatures simply cowered in a corner as their eyes and lungs burned. Many were mothers who managed to escape while their babies could not, leaving them abandoned—the sort of fate that had put Mandy in a cage at a raccoon giveaway.

The company was not an outlier in its practices; that was how the industry worked. "It was very inhumane," Gates recalls now. "That was a push to for me to develop techniques that were 100 percent humane." And that is how, 41 years later, he came to be standing atop a two-story house on a brisk March morning in suburban Markham, Ontario, where he and his daughter Cassandra arrived behind the wheel of a white van emblazoned with the name of his business, AAA Gates' Wildlife Control, a raccoon logo, and their motto: "The Animal's Choice."

The home's owners had been hearing noises; Gates was their choice, not because of the old "AAA" phone-book trick, but because of his reputation. He's widely recognized as a pioneer, even a revolutionary in peacefully handling domestic tensions with urban wildlife—what might also be termed pest control, although he doesn't much use the term himself. There are no fewer than eleven vans in his fleet, as well as a franchise in Vancouver, British Columbia, all testifying to his success and to a flourishing of public compassion. To a great many people, even a so-called pest, an animal who trespasses into one's home, who may pose a threat to property and personal safety, is someone to treat with kindness.

One might understandably wonder: is that really *important*? It's nicer than the alternative, certainly—but when it comes to considering wild animals as fellow persons, wouldn't these animals come well down the list of candidates? Heck, the creatures for whom people like Gates care are not always considered to be properly wild animals, or even parts of nature. And with species dimming like stars on a hazy night lit up by streetlamps, does the well-being of rats and raccoons and pigeons matter all that much? They're certainly in no danger of going extinct. Yet it's in them that the reality of wild animals in urbanized places meets the ideas we have about nature. Our relationships to them are definitional.

There is an unexamined tension in how wildlife is simultaneously welcomed in cities yet expected to remain segregated. While some people delight in a groundhog under the deck or sparrows in the eaves, others see an invasion. Meanwhile most every large conservation organization now devotes at least some of their attention to nature in cities—but they're celebrating biodiversity, beauty, and amenity, not necessarily offering practical guidelines for coexistence, particularly when things get messy. They're not asking what it means for urban nonhumans to be accepted as members of a shared society.

"For those who feel that how wild animals are treated is a morally relevant concern, an ethic for urban wildlife becomes important," writes John Hadidian, the now-retired former director of the Humane Society of the United States's Urban Wildlife Program. Gates, whom Hadidian credits with developing techniques that the Humane Society now works to popularize, embodies one possible ethic: that animals, in Hadidian's words, "all deserve moral consideration and humane treatment," even when they are considered, or become, pests.

Like weed, pest is not a category with a biological basis, but one defined by human perception: they are unwanted and bothersome. Typically they receive little legal protection. In Ontario, where Gates lives, almost any animal can be killed for causing property damage, which can be as trivial as a skunk's digging for grubs in the lawn or a pigeon pooping on a windowsill. In New York, an animal doesn't even need to cause damage; it's enough be judged a nuisance. In many places most wild animals are so-called pests, or species for whom that label is only a few episodes of human inconvenience away. It can apply not only to the usual suspects, like rats, or boundary-crossing species like raccoons, but even those who are charismatic, like bald eagles in an Alaskan fishing village, where a bounty of scraps transmuted them from beloved icons to messy encumbrance.

A great many lives are affected by how we deal with animals who bother us. And beyond their sheer numbers, the predicament of creatures with whom we share urban habitats could be seen as emblematic of life on a planet where humans are everywhere. The ethics of cities and so-called pests and unwanted wildlife are central to the Anthropocene. Will it be an era in which wild animals are accepted only if they cause no inconvenience? Where they are welcome so long as they stay in designated spaces, making no contact or ethical demands upon us? In which, depending on how we categorize them, some lives have value and others none? In cities, we are presented with what Dalhousie University environmental studies professor Erin Luther frames as nothing less than the challenge of "what it really means to live with others in the world."

◆　◆　◆

WHEN GATES AND CASSANDRA ENTER THE ATTIC, clad in jumpsuits and facemasks, it must be quite the shock to the raccoon mother and babies inside. It's probably the first time in their lives that *anyone* has entered, and now headlamps blaze through their cocoon of darkness. The babies mewl in sudden discontent as mom retreats into a maze of chipboard and two-by-four joists where the roof slopes to the floor.

"Hey mama," Gates says in a measured, level voice—which is how he sounds in conversation, too, and also how he moves. Calmness is both something he wants to convey to the animals and a job requirement: when, say you're 15 feet up on a ladder with a mother squirrel somewhere inside a hole several inches from your face, you need to keep your wits. Here there's not much to worry about, though. Raccoons can certainly be ferocious in self-defense, but mom simply hides.

Before calling for help, the home's owner looked fruitlessly for signs of how the raccoon got inside. When Gates and Cassandra arrived, they set up their ladder, climbed to the roof, and went straight for the ventilation pipe. Builders often cut a larger-than-necessary hole, providing some wiggle room for the pipe's final positioning, then seal the gap with a sheet of neoprene. It's a weak spot in an otherwise impenetrable barrier of plywood and roofing felt and shingles. If a raccoon can get her hands beneath the neoprene and dislodge a few nails, the attic awaits.

Gates describes raccoons as miniature home inspectors. The same goes for squirrels, who specialize in drainpipes and in soffits, those little flaps of plastic or aluminum beneath the overhang of a roof. (Full disclosure: I did not even know what a soffit was until Gates explained about squirrels. It's interesting to think that many urban animals know more about buildings than the humans inside them.) They need only find enough of a gap to stick in a muzzle and start chewing; hours or days later, they'll be able to fit their body through.

"They will show you where the flaws exist," Gates says. Often he finds signs, such as a peeled-back piece of metal, of a raccoon's exploratory probings. Lessons spread between raccoons: in a neighborhood where one has learned to open a particular kind of vent, others soon follow suit. He thinks cultural differences exist between the area's raccoons, with some specializing in those sturdy, elegant homes built at the turn of the twentieth century and others in more recent construction styles. Modern plastic roof vents have been great for business.

Markham's raccoons also know a lot about insulation. Most of the attic floor is covered by a fine cellulose powder, made from ground-up newspaper, that leaves fewer voids than fiberglass. It must irritate their eyes and noses; when a raccoon gets into an attic, Gates says they seek out inclined areas where the cellulose is mixed with fiberglass to prevent it sliding like snow. It's in precisely one of those sloping spaces, back in a corner that provides some protection, that mom made a nest. Inside are the babies, six of them, with downy hair, closed eyes, and large heads, pressed into a ball of natal warmth, keening in high chattering voices at this invasion of the security that until now has been their entire world.

Back when he started, Gates says, most people equated killing animals with solving problems. By those standards the formaldehyde-fogging machines actually represented progress. Even companies that promised humane methods didn't always see them through. They might set a trap and not check it, instead relying on clients who, busy with their own lives, would often forget, leaving trapped animals to die slowly. If they did check, it was common to simply catch a mother and unwittingly leave her babies to starve.

This sometimes still happens—and, now as then, some homeowners take matters into their own hands. John Griffin, who succeeded Hadidian at the Humane Society and headed a program they ran to demonstrate animal-friendly ways of handling wildlife issues, told me about what he saw. People would try to smoke animals out of chimneys, which is even worse than formaldehyde, or set off pesticide bombs to rid attics of bats, or seal animals into walls. Each summer, he said, wildlife rehabbers are still overwhelmed by babies orphaned when people took their moms away.

Once Gates took a call from a high school science teacher who escorted him to a room below the attic where a rifle was propped in the corner. There were no fewer than 15 bullet holes in the ceiling. The teacher turned his roof into a sieve but never hit the raccoon. Those sorts of calls are rare now; it's far more common for someone to simply buy a live trap—they're easily available at most any big-box hardware store or agricultural supply center—and catch the animal themselves. Usually they drive the poor creature out into the countryside, thinking they're doing a favor, but it's close to a death sentence. It's hard enough staying alive in an unfamiliar place, but putting a city raccoon in the country isn't much different than plucking one from the deep forest and dropping him off downtown. Indeed the difficulty of relocation has led some people to consider a quick death to be a more humane option.

The decision of whether to kill an animal is usually left to property owners or the people they hire, although laws sometimes require it. Several years ago, the state of Indiana tried unsuccessfully to mandate that wildlife removers kill every raccoon, possum, and coyote they caught—a proposal that many in the business fought, but others welcomed. In Pittsburgh, Pennsylvania, where the municipal animal control department loans traps to citizens, every groundhog, raccoon, and skunk captured this way is euthanized. The city reported killing 1,500 raccoons yearly from 2011 to 2014. New York City likewise requires municipally trapped raccoons to be killed as potential rabies vectors. At the other end of the spectrum, Washington, DC passed a law in 2010 requiring nonlethal resolution to animal conflicts—but it applied only to commercial operators, leaving homeowners and property managers to act as they please.

In short, urban wildlife conflict is guided by a hodgepodge of practices and regulations that display relatively little concern for animal well-being. John Hadidian thinks this reflects the relatively young field's cultural DNA: the rat-catching businesses that first emerged in European cities and migrated to cities in North America, government pest control services offered to farmers in the nineteenth century, and wildlife management agencies that evolved to regulate hunting. As urban sprawl brought more people and wild animals together in the twentieth century, local governments simply referred citizens with wildlife problems to trappers. They evolved into specialized wildlife removal services, while the existing pest control industry expanded to handle creatures who now occupied a gray space between wildlife and pest. Profitability, rather than consideration for the animals themselves, dictated their treatment.

A few decades ago that wasn't a problem—but the culture of cities and of the public at large has shifted. Social scientists have documented how, in their argot, people in the United States are becoming oriented to mutualism rather than domination; they value coexistence and take a live-and-let-live approach to wild animals. That's not always true, of course—either because some people don't much like animals or because even the most animal-loving person can waver after several sleepless nights of having their imaginations inflamed by gnawing sounds from the walls. Sometimes Gates gets calls from people who just want an animal gone and don't care how it happens. Those are much rarer than they used to be, though. He also notes that many of his customers are families with children. It's harder to rationalize killing when confronted by a child's unflinching compassion.

Cassandra gently places the baby raccoons in a canvas sack. They'll go in an insulated box on the roof. Mom will leave when she's ready and, if all goes according to plan, she'll find them and move them to a new nest, after which her hole will be sealed. It won't be much inconvenience. She likely has dozens of sites lined up. When John Hadidian radio-tagged raccoons living at the edge of Rock Creek Park, that ribbon of ancient forest running through the heart of Washington, DC, he found one female who made the rounds of some 200 den sites: tree hollows, coppices, well-positioned logs, sewers, and a few attics.

Gates and Cassandra take the raccoons downstairs and briefly display them for photographs and exclamations of cuteness. Then Gates plugs an extra-long extension cord into an outdoor outlet and climbs back to the roof. The cord is for a heating pad that, along with an old sweatshirt and some slabs of Styrofoam insulation, goes inside the box, which they made themselves and resembles a cooler with a sliding door. The whole setup is one of his innovations. Without it the babies would likely freeze. They go in last, quivering and clinging to one another. It's a heart-rending scene, but long experience taught Gates that this works best. The babies burrow into the sweatshirt as he closes the box and screws it securely into the roof.

Job done, he pauses for a moment and takes in the panorama of rooftops and backyard trees stretching in every direction—a per-spective shared, when one thinks about it, not by the residents of those homes, who rarely have reason to climb atop them, but by birds, squirrels, and other neighborhood creatures. (Raccoons belong on that list too, obviously, but nearsightedness denies them the view.) Customers often believe, Gates says, that the prolifera-tion of urban wildlife results from habitat destruction that pushed them into cities. But while forests were certainly cleared to build this suburb, and it's inhospitable to many species who once lived here, for those who can adapt it's now a home. *This* is their habitat.

"Animals deserve to be in our urban environment as much as we do," Gates says. His words are deceptively simple: not only is the city their habitat; they *belong* here. It's as much a social as an ecological judgment.

This customer had no qualms about letting their unwanted houseguests stay in the neighborhood, but some need convincing. Sometimes they also need reassurance that it's worth paying for the time it takes him to make a detailed inspection of their home, looking for potential entry points and animal-proofing them— something that's now a standard part of his package, reflecting an approach that embraces prevention rather than cycles of temporary resolution and future conflict. This seems like common sense, yet it's not a universal practice. It's still common for pest controllers and wildlife removers to offer a cursory inspection and move on to the next job. In one recent survey of towns in the United Kingdom that provide pest control for their residents, for example, only half gave advice on exclusion and deterrence.

To Gates the inspections are also about creating a culture of coexistence. If he leaves the roof vents unscreened and a raccoon returns, what ended today in coos and magnanimity could fester. Tolerance dwindles when problems recur. "If I can provide a 100 percent animal-proof solution moving forward, then they are no longer bothered," he says. "It teaches them that we can live in harmony."

◆ ◆ ◆

IT'S ONLY APPROPRIATE THAT RACCOONS HAVE FIG-ured so prominently in Gates' life. As we saw earlier, they occupy a unique position in North America's cultural history, at home in both wilderness and human settlements. Many Indigenous tales feature raccoons; a favorite of mine involves one who ate the meals of two old, blind Menominee men in a lakeside village, causing them to bicker, then untied the knots in a rope they used as a guide to shore. Realizing what he'd done, the raccoon set about retying the knots, whereupon the men found him. As they stroked his fur, they forgot their enmity and the resentments of age.

Underlying the narrative is the intelligence that makes raccoons so delightful and sometimes exasperating. In Toronto, where so-called trash pandas are the city's unofficial mascots, the municipal government spent $31 million developing raccoon-proof bins for compostable waste. They hadn't even finished distributing the bins when reports started arriving of raccoons who'd figured them out. (The manufacturer had stressed that the bins were raccoon-*resistant*, not raccoon-*proof*. Promises were for politicians, not people who truly knew the animals.)

Gates' own favorite story involves a raccoon living in a garage. His inspection revealed the building to be structurally sound. He couldn't figure out how she kept entering until, as he climbed up into the rafters, he saw her climb down a column on which the electric garage door opener was mounted, pause, push the button, and saunter out. She had a nest with babies, meaning she had used the garage for months. As best as he could figure, she would go outside at night while the homeowner slept, then close the door when she returned in the morning's wee hours, leaving her humans none the wiser.

Even as adversaries raccoons merit a certain respect. But what about those creatures to whom we're not so predisposed—namely rodents, particularly rats and mice? The line between wildlife removal and pest control is blurry, but certainly many people would put them in a fundamentally different category from raccoons. To be sure, people are still sometimes kind to them; humane mousetraps have thousands of reviews on Amazon. When I asked Margaret Robinson, the Mi'kmaq scholar, about how traditional values of respect and reciprocity and kinship applied to the so-called pests in her life, she told me of how she once caught mice in her apartment and put them in a Habitrail enclosure for pet hamsters. They chewed their way out after hiding the hole with bedding. ("It was like *Escape from Alcatraz*," she said. "I concluded that these are independent people.") She decided not to worry and the problem resolved itself, but Robinson could appreciate how this let-do attitude wouldn't serve in a true infestation.

And that's for mice, creatures who figure sympathetically in many tales—they're Cinderella's friends, after all. They're not nearly so objectionable to most people as rats, specifically *Rattus norvegicus*, the misnamed Norwegian brown rat, who originated in southern China and is perhaps the most-despised and most-killed free-living vertebrate on Earth. Black plastic boxes full of poison baits are so ubiquitous in cities as to escape notice. The notion of protecting rats is beyond the pale. Fears that Washington, DC's humane animal conflict laws applied to rats and mice drew national attention: Rush Limbaugh even chimed in, warning that "it shows what happens when you get hold of these young kids in kindergarten on up and you start inculcating them" with liberal ideas. The furor subsided only when DC City Councilmember Mary Cheh insisted that the law applied to "wildlife—not rats or mice, but wildlife." Even when rodenticides are banned, it's not to protect rats, or out of sensitivity to the excruciating process of dying from internal bleeding over a period of several days, but for the sake of hawks and other charismatic predators.

Yet by any reasonable ethical standard, rats deserve some serious consideration. Research abounds with empirical insights into the richness of their intelligence and emotions, which we touched on earlier. They laugh when tickled, feel regret, plan for the future, share food—and do so more readily with rats who are hungry—take care of one another's babies, and jump for joy. (The word for the latter is the delightful *freudensprung*.) Experiments on rat empathy have received much attention, both because of public fascination with rats—the flip side of persecution, which sometimes yields a certain begrudging affection—and the possibility that they illuminate something essential about the human condition, too. "The Kernel of Human (or Rodent) Kindness," read one *New York Times* headline.

Much research also exists on urban rat ecology and how the evolutionary pressures of urban life may have made them smarter and more empathetic. Yet even that treatment of rats as subjects of scientific study doesn't capture the perspective of people who *know* them—people like Maria Chen, a student at the University of British Columbia's animal welfare program, who had been recommended to me by the animal law attorney Victoria Shroff, whom Chen had greeted at a campus meeting while wearing a scarf in which her two rats snuggled. Chen introduced me to one of them on a Zoom call: a chunky gray fellow named Beastie, whom she'd adopted from a Vancouver animal shelter.

Beastie perched on her shoulder, nuzzling her ear. He loves having his belly and the bottom of his chin rubbed, and after a while will twitch his dignified nose and whiskers in satisfaction. When he's quite happy he grinds his teeth, said Chen, and when he's *really* happy, when the world is so good that he just overflows, he engages in a behavior known as boggling: grinding his teeth so intensely that his clenching jaw muscles cause his eyes to pop in and out. I had never heard of this. Neither have many people who work with rats in labs, Chen said—their rats are never completely relaxed. (Rats and mice do not qualify for protection under the US Animal Welfare Act, which governs the treatment of captive animals. Even the minimal consideration of giving them sufficient room to move is too much to ask.)

Chen said that Beastie is much more sociable than Maggie, who passed away in 2019, three years after she and her partner adopted her. Three years is a ripe old age for a rat, and she might have been much older, as she was already an adult when they found her at the shelter. Maggie bonded quickly with Chen's dog Pudding, but it took months before she finally climbed to Chen's shoulder and, in the ultimate sign of trust, fell asleep. Chen would take her to school and out on errands. She enjoys seeing how people react. "Sometimes people look at a rat and they're not even scared," she said. "They're like, 'Oh, is it a mouse? Is it a hamster?' Then immediately when they realize they've been looking at a rat, they kind of freak out." Reputation precedes reality.

Of course domestic rats are not identical to wild rats; they've been bred to be calmer and less aggressive. Some have likened the difference to that between dogs and wolves, by which token Beastie would be like a husky, resembling a wolf but not acting like one. Much as the traits humans celebrate in dogs, especially the capacities for love and loyalty, are magnified by domestication but present in the wild ancestor, so it is with rats. And while taking wild rats as pets is absolutely discouraged, some people who do so report that they're quite similar to their domestic brethren. The affection, playfulness, and individuality Chen loves in her rodent friends are fundamental qualities of rat-ness.

Rats are an interesting species with whom to think through ongoing debates over similarities and differences between humans and other animals and to consider whether these are morally meaningful. Maybe the pleasure rats take in a good belly scratch is a more meaningful similarity than any difference. And while extensive protections are unrealistic, simply taking the steps necessary not to attract them, and therefore to avoid the dilemma of killing them, seems reasonable. Yet even that is a struggle. "It just amazes me that these are really common-sensical simple things," said Bobby Corrigan, probably the world's best-known rodent control specialist, who when I called was working on the second edition of *Rodent Control: A Practical Guide for Pest Management Professionals*. "But when it comes to vermin, most people turn it off. They're like, who cares?" Again and again he tells clients to secure their trash. Again and again he returns a few months later to find that nothing has changed. He called this "the paradox of *Homo sapiens*. We're not really sapient"—wise—"at all."

◆　◆　◆

I WASN'T SURPRISED BY CORRIGAN'S ASSESSMENT. I've heard it many times. The vehemence and depth of his frustration, though, was unexpected. He is a rat catcher who doesn't want to catch rats. Corrigan reminisced about one who eluded him for weeks on the commodities trading floor of the former World Trade Center, brazenly venturing out at lunchtime to pick up unguarded slices of pizza but somehow keeping the location of his den secret. Corrigan paid his respects before killing him—something he felt pressured to do, what with his clients' expectations and the trading floor's testosterone-charged air, and now regrets. At the end he didn't want to find the rat at all. "How can you not respect an animal that intelligent?" he said. Given his occupation, this might seem paradoxical, but Corrigan has an ethos.

He sees some killing as unavoidable. If there are hundreds of rats in a busy park, or an infestation of mice in a restaurant, they're not going to be live trapped. It's too time-consuming, expensive, and logistically challenging. This is sad, but nature is a harsh place. "I would say, to anyone who says you shouldn't kill *anything*: How are you going to do that? Give me some real, practical examples," he said. But killing should only happen when it's truly necessary, and "if I have a choice to save something without killing it, I'm going that route every time." He despises the mentality of people who reach for a rifle at the sight of a groundhog and grows angry at the thought of hunting for fun. And when killing happens, he wants it to be as painless as possible: not poison, but a suffusion of carbon dioxide into a rat's burrow, so that death comes like sleep. Then it's time to take care of the trash.

"An uneasy truce based on parallel planes of existence" may be the best of all realistic outcomes for rats and other rodents who thrive in humanity's shadows, writes environmental theorist Christian Hunold, but it would be vastly better than what prevails now. Within that truce can be bright spots. One of Corrigan's favorite colleagues is Rebecca Dmytryk, co-owner with her husband Duane Titus of the Moss Landing, California-based Humane Wildlife Control. Dmytryk started as an animal control officer in Los Angeles in the mid-1980s; then she founded the California Wildlife Center, a hospital she ran for several years before specializing, in that land of oil spills, earthquakes, and wildfires, in rescuing animals during disasters. About a decade ago the sheer volume of calls she received about nuisance wildlife convinced her to start a business. She started with an explicitly no-poison, no-kill ethos. She found more sympathy than one might expect. "Everybody in their circle hates rats," she says of her clients. "Then they come to me, and they get the approval that it's okay not to hate them."

Like Brad Gates, whom she called "the grandaddy of us," Dmytryk emphasizes the importance of keeping animals out in the first place. She and Titus conduct detailed inspections of building exteriors, sealing every crack, then live trapping the rodents and releasing them outside—not only because it's ethical, but because it's the best way to test their work. They set up night vision cameras to watch how the animals try to re-enter; if they keep getting back inside, they use nontoxic fluorescent dyes into which rodents' feet and paws can be dipped, leaving trails back into the house that glow under ultraviolet light, and surveillance cameras, controlled by smart phones, to pinpoint the entries. Often her clients start sending cute photos. "They're getting engaged with animals they once considered awful and evil," she said. "I truly believe this: that given a choice, people would rather not kill an animal. They just want their problems resolved."

Dmytryk is especially frustrated with the mentality of many agricultural producers. In California, extra-lethal rodenticides, so potent that a few grains can turn a rodent's innards to mush and that are known to kill large numbers of nontarget animals, are banned from residential use. But on farms, which cover ten times more land than do cities in the state, their use is still allowed. "It's easier for them to just throw poison" at their problems, she said, and the mere presence of wildlife may be considered a problem. The other day someone had left a message about needing help with a family of foxes. "I called the guy up," Dmytryk recalled. "He's like 'Yeah, we got some foxes here in this field.' I go, 'What are they doing?' He said, 'Well, they're just here. And they're running around.' So then it turns out that it's an ag field. I go, 'Okay. Do you have a fence around your field?'" He did not. She turned down the job but is sure that he found someone else to do it. "And that"— the moral shadow of the foxes' slow, avoidable deaths—"is on your spinach," she said.

"It's this anthropocentric point of view," Dmytryk continued, "instead of ecocentric"—which is to say, respecting the intrinsic value of Earth's ecosystems. That's how Dmytryk understands what she does: compassion for each mouse and fox is part of what it means to be a good planetary steward. Ecological collapse is imminent, she thinks, "and I'm just kind of resigned to that. Now I'm focused on expressing humanity to other beings." This could sound fatalistic, even nihilistic, yet she had recently finished tests of a rodent-proof fencing system she invented. After we talked, she sent a video of a rat trying futilely to jump over it. "We land on the moon. We go flying around Mars. We can build a fucking rodent-proof barrier, okay?" she said. "Take responsibility. Be a good steward."

Farmers and ranchers, Dmytryk said, should be required by law to protect their fields and livestock before being allowed to kill wildlife. No poison before fences. To her it's a basic principle, like keeping pesticides out of drinking water. It's vision that's morally consistent and sensible—and, at the present moment, almost impossible to imagine. The backlash would be swift; one can envision culture-war fallout in which urban elites are accused of bankrupting farmers with their rat-hugging demands. Such a change would require broad public support and a sea change in cultural attitudes toward rodents.

◆ ◆ ◆

AND YET—SEAS CAN CHANGE. WITNESS THE EXAMPLE of Narda Nelson, a pedagogist—a specialist in the practice of teaching—and proponent of what's known as the common worlds approach to early childhood education. The name comes from French philosopher Bruno Latour's term for entangled communities of humans and nonhumans, with a nod to theorist Donna Haraway, who invites people to reconsider the word *world*: not as a noun but a verb, something constructed through the relationships we build. It's premised on the idea that crises of extinction and ecological collapse stem from a belief in human supremacy—a belief so widespread that values of diversity and justice are generally understood as applying only to humans, and even then unevenly so, with white Europeans and their descendants tending to end up on top. The common worlds approach rejects this. It calls for relations founded on ethics of care, on a sense of kinship with creatures who are kin in deep evolutionary time and in present circumstance.

All that might seem like a lot of conceptual baggage for kids still watching *SpongeBob SquarePants*. But in practice, Nelson explained, it's much simpler. It involves encouraging attentiveness and empathy to the animals in everyday life: worms in a classroom compost bin, the deer and banana slugs they meet on walks to the forest, slime molds, trees about to be cut to make way for a sanitation station. Educators encourage children to imagine their experience of the world and the relationships in their lives. The goal is not to learn species names or enhance cognitive development; rather, it's nourishing a sensibility, an awareness. "The children do not call things *it*," Nelson says. "It's a *him, her,* or *they*."

So it was that, one March morning, a group of children from the University of Victoria, British Columbia's early childhood education center happened upon a rat. This wasn't part of the plan. They and their teachers were walking to a forested spot on the campus to look for animal tracks. But there was the rat, right on the sidewalk, on her side, convulsing in agony. It was obvious what was happening, and Nelson thought she would keep the kids moving rather than dwell on something so awful. "It was the children who would not let it go," she said. One of them wondered whether they could give the rat a Band-Aid. Another interpreted the rat's shaking as cold; maybe they could put leaves on her, for insulation. Someone suggested that she was a grandmother trying to get home. Finally one of the children said that she must have been poisoned.

In the following months the rat kept coming up in conversation. Not long afterward, a child brought an urn containing her beloved cat Rosie's ashes to show-and-tell; she talked of Rosie's life and how, when she fell ill, they had to take her to the veterinarian to die. One student mentioned how their grandfather died recently, too, and had been cremated; the students wondered whether they should have done that for the rat on the sidewalk. Another asked: Why don't we do that for *every* creature who dies?

It was a deceptively simple question, touching on life's harsh realities—the constraints of practicality, the fact that some lives are valued and others are not—and Nelson was saved from an uncomfortable answer when someone asked to see Rosie's ashes. "Demanding to know why the rat was excluded from this type of care lays bare the crux of the matter," she would later write. "What does 'ethical inclusion' look like with a creature few want to live with?"

To even entertain the possibility of caring for dying rats, Nelson told me, was powerful. It disrupted the way all lives are not seen as equal, and how some deaths are not considered worth grieving. It upset the categories of disease carriers and vermin into which rats are assigned by default, and it pushed the students to be aware of the industrial-scale poisoning with which—whether they knew it or not— their own lives were intertwined. It challenged the ethical divide between humans and rodents. "We're living with them always," Nelson said. "They're not the charismatic ones. They're not the ones we necessarily want to see. But how do we live with them? Because we have to anyways."

The children did not organize to demand change at the school, nor were they encouraged to. That, said Nelson, is not what common worlds is about. The teachers are there to nourish conversation, empathy, and imagination. At least the rat's ordeal was witnessed. Later Nelson wished she'd put the poor rat out of her misery, and she did talk to one of the school's groundkeepers, who said simply that he couldn't control where rats died. But how extraordinary it was for teachers even to think these things through, to ask what education in a more than human world means and what it means to question narratives of human supremacy, and to honor a dying rat by asking who lives and dies well in the world we have made. Who knows what will grow from the seeds planted by that simple, radical act.

Years later, when a nearby construction project drove rats from their homes, setting off a small panic among some segments of the education center, the students in Nelson's class talked it over and decided against trying to trap and relocate them. That would break up families and consign some to death. Instead they decided to let their playground be a safe passage. They'd avoid rats if they met them, be careful with their food, and keep an eye out for droppings. Every so often they did see rats. There was one with light brown fur whom they came to know by sight, and they took to calling her Rosie.

◆ ◆ ◆

BACK IN TORONTO, AFTER LEAVING THE BABIES, GATES attended to the day's other calls. Baby raccoon season was in full swing; that overlaps with baby squirrel season, followed by baby skunks, after which come baby birds, and then a second flush of squirrels in summer. His company's eleven trucks operate year-round. The only time things are ever truly quiet is during especially cold, snowy winters, when raccoons decide to conserve their energy and enter torpor—a state of reduced metabolic activity that falls short of hibernation's dormancy, kind of like a month-long nap—and even squirrels, whose activity can be relied upon for year-round business, grow quiet in their nests.

A few weeks earlier he removed a squirrel family from above a doorway where mom took advantage of some sloppy woodwork around a drainpipe. In order to reach the babies Cassandra had to cut into the structure with a jigsaw. As she did, and then as they were transported onto the ground, mom hung back, gnawing on a black walnut. Gates had never seen this before. He didn't think it had anything to do with being hungry. Instead it struck him as a stress behavior, the expression of a frantic mother watching powerlessly as giants took her children. Fortunately, she took them all back as Gates and Cassandra stepped away, picking each one up by a hind leg—a trick to make them curl up for easy transport—and dashing across the fence tops to some other den before returning for the next.

The baby raccoons are not so lucky. When Gates and Cassandra return the next morning, they find that mom has only returned for one. The rest mewl with fright and hunger, but he knows from experience that trying to feed them will sicken stomachs that have known only the nutrients and microbes of their mother's milk. The next morning they're still there. Gates doesn't think this is because mom doesn't want her babies. Instead he thinks the incident was simply too traumatic for her. It's rare that this happens, but it does. He does all he can to avoid hurting animals, but it's not in his power, or anyone's power in this world, to prevent every wound. He can only try his best to heal them, and not all will mend neatly.

But the maternal embrace of a raccoon mother is not confined to her own babies. So long as they're roughly the same age as her litter, a raccoon mom will readily adopt others. Gates and Cassandra unscrew the box and take the babies down to their truck. Fortunately, they've another raccoon job nearby later in the day, and after capturing those babies they combine the two litters. When they come back a day later, all the babies are gone.

8

The Caring City

ONE HUMID SPRING EVENING ON A BUSY PATH NEAR Washington, DC, I spotted a pair of baby groundhogs: two furry little bundles of vulnerability, one injured and unable to walk, the other wobbling blindly through short trailside grass. They belonged in their mother's burrow, with her care and several feet of earth separating them from worldly concerns, but she was nowhere in sight. Perhaps she'd been killed by a coyote, an owl, or even an off-leash dog. The babies' fate was in my hands. Would I continue on my way and leave them to die or try to help?

An argument could be made for letting nature take its course. The groundhogs would become meals for others, and though the cycles of life are painful, Earth would be barren without them. Intellectually I knew and accepted this—and, if I'm being entirely honest, I didn't want the obligation and uncertainty that suddenly intruded into a quiet evening's plans. But suffering is hard to ignore when it's right in front of you, and helplessness even more so.

I recalled reading about a local wildlife hospital and looked up their number on my phone. They were already closed but someone answered and offered to help if I brought the groundhogs. I had nothing to carry them—I had been on a run—but a passing cyclist stopped and, in a stroke of good fortune, turned out to be headed home from work as a physical therapist. She pulled a set of scrubs from her backpack; I gathered the groundhogs, too weak to flee, into the shirt, then called an Uber. Would I even be allowed into the car, I wondered? I didn't need to worry. The driver was a young woman named Fantazzia and she was thrilled to help, slashing through rush-hour traffic while talking of how much she and her young daughter loved the deer in a pocket forest behind their apartment complex.

Thus we arrived at City Wildlife—opened in 2013 to care for the District of Columbia's sick and injured nonhuman residents—where a volunteer took the babies and promised to call. That night I heard from Paula Goldberg, the hospital's executive director. The groundhogs were in bad shape indeed. It had been days since they nursed, and infections overwhelmed their infant immune systems. One had likely been struck by a bike and could no longer move his hind legs; he had to be euthanized. Goldberg took the other home and cared for her through the night. She survived for several days before succumbing. It was a sad moment— more than I expected, as is so often the way with animals—and a clarifying one.

I was a nature-loving, conservation-minded person; conservation had taught me to appreciate biodiversity, protect habitat, recognize the benefits that ecosystems produce, and know the identity and life history of groundhogs. But at that moment I needed help for two specific wild animals—not populations, not species, but beings trembling in my hands—and no conservation organization or land trust or state wildlife agency could offer much help. Neither, for that matter, could a veterinarian, as they tend only to domestic animals. Into that gap stepped City Wildlife. I asked Goldberg if I might spend a day there.

Several weeks later, when peak baby squirrel season passed, she invited me over. I arrived around nine on a morning already hot enough to stifle the birds flitting through community gardens and a railroad-corridor vinescape beside the hospital, which is located on a nondescript commercial block in the city's northeastern quadrant. In the parking lot were cages, draped with sheets against the sun, containing three young house sparrows, two starlings, two mourning doves, a pair of opossums, and a squirrel named Houdini. In an examination room inside, Goldberg and Eliza Burbank, the hospital's veterinary assistant, attended to the day's first patients: a dehydrated juvenile crow found grounded at a dog park and a fierce-eyed mockingbird with one foot tangled in string.

One of the mockingbird's toes was broken. His leg also had a puncture wound and several torn tailfeathers. A cat attack, most likely, and over his vocal protests Burbank administered a course of antibiotics and painkillers. The crow got a shot of fluids and was placed with another young crow recuperating in an incubator. As if on cue they cawed in unison. I tried to recall what I knew about communication in crows, a species so very renowned for their intelligence and sociality. Kevin McGowan, the Cornell University ornithologist who studied crows, once told me that, over his three decades of observing crows, he learned to make rudimentary translations of their vocalizations—their language, some would say—and that it mostly seemed to involve food, predators, and their surroundings. Their conversations were hard to follow, however, because slight changes in intonation altered the meanings of calls. He likened his struggles to a native English speaker listening to Chinese. He also mentioned that juvenile crows could often be found reciting vocalizations to themselves, as if practicing.

What might these unusually brainy birds, plucked from their homes and surrounded by giant strangers, be saying? Burbank, at least, knew a little of how they communicate. "When they gape like that," she explained, "they're asking to be fed."

◆ ◆ ◆

AMONG THE FIELD GUIDES AND CARE MANUALS ON the bookshelf above Goldberg's desk was a well-used copy of *Answering the Call of the Wild: A Hotline Operator's Guide to Helping People and Wildlife*, written by Erin Luther, the environmental studies professor whose theories about coexisting with unwanted urban wildlife and so-called pests we encountered in the last chapter. Before entering academia, Luther spent 14 years taking calls on the Toronto Wildlife Centre's hotline, which receives some 30,000 queries each year: about orphaned or injured animals, potential conflicts, seemingly odd behaviors, and most anything else one could imagine involving wild animals in the city.

Luther came to understand her experiences, all those thousands of calls, as part of something larger: people grappling, at a moment of ecological, cultural, and scientific upheaval, with their ethical relations to animals. "People were looking for guidance," she told me. "What might right relations with nature look like in the city? What are our obligations and duties of care? We're coming into this newly defined time of the Anthropocene, and we're thinking more about how we live with wildlife: not just preserving those populations, but what our relationship to them is."

Compassionate approaches to wildlife conflict and so-called pests are one part of this. At their heart are what ethicists call negative duties: things we should not do. If we understand other animals as thinking, feeling, self-aware beings, with friends and families and relationships, as fellow persons and neighbors in a more-than-human city, then we ought not to kill them unless truly necessary. Eating one's tomatoes or sneaking into an attic isn't grounds for exile. But what about positive duties, or what we *should* do for others? Not necessarily as a matter of legal obligation, but because deep down we know it's right?

Trying to help the orphaned baby groundhogs was one example of a positive duty. And though most wildlife rehabilitators work from home, doing their best on a shoestring budget in a spare room or, if they're fortunate, a shed, I happened to be near a fully-equipped wildlife hospital.* By the standards of human hospitals, of course, City Wildlife was basic—including Goldberg, there were four full-time employees, and they would have been overwhelmed without volunteers—but it was more than all but a few cities have. City Wildlife hints at what is possible when caring for wild animals becomes part of a community's institutional fabric.

The hospital could even be understood as embodying the concepts of Sue Donaldson and Will Kymlicka's *Zoopolis*: animal citizens and denizens are not entitled to the full suite of medical care guaranteed to a city's human residents, but they still deserve something. About half of City Wildlife's patients come from the city's Animal Care and Control department. In many other cities, people with wild animal problems are directed to pest control companies or receive little help at all. In Washington, the baseline is more compassionate.

As Burbank finished with the crows, Goldberg headed to a back room and instructed interns on the finer points of handling several dozen hyperactive mallard ducklings. A former physician's assistant, she was introduced to wildlife rehab when, while volunteering at the Smithsonian Museum of Natural History, a colleague smuggled into work three orphaned opossums whom she was nursing to health. Nowadays Goldberg still has an old-school doctor's-office air: slightly frazzled and given to poking fun at her

---

* Consider Rachel Parsons, a wildlife rehabilitator in eastern Maine who once helped me treat a fox and a groundhog—Robin and Arthur, as I knew them—for mange. (The treatment involved an antiparasitic drug injected into, respectively, meatballs and carrots, which I left where they would find them, monitoring the results with trail cameras and patience.) Parsons' rehab center is a well-equipped shed beside her country home. At times, such as baby squirrel season or when mouse pups need feeding every two hours, she works around-the-clock. Cash-strapped and understaffed as City Wildlife is, rehabbers like Parsons, of whom there are thousands, have an even harder time.

own absent-mindedness, with loosely tied shoelaces that threaten to trip her up but never do. She's calm under pressure and what needs doing gets done. After being weighed, the ducklings are taken to a kiddie pool outside. They frolic like kids at a water park.

Back inside, the day's next patients had arrived: three tiny opossums delivered by a woman who, while walking her dog near Capitol Hill, saw a car hit their mother and had the presence of mind to check her pouch. (Opossums are America's only marsupials, that mammalian order known best for kangaroos.) Hairless and pink, eyes still shut, the babies were so young they likely wouldn't survive. They were also dehydrated and one, unresponsive to pinches, appeared to have a spinal injury. He'd probably be euthanized immediately. For this task the hospital uses a benchtop chamber into which anesthetic gas is pumped. Death is supposed to come quickly and painlessly. "It's less awful than it would have been, slowly starving to death," Burbank said. "It's still a service to them."

Opossums are easygoing, slow-moving creatures who almost never attack humans. When threatened they may hiss and bare their teeth, but if that fails they pretend to be dead. Still, perhaps because of their appearance, resembling giant rats to an undiscriminating eye, people often treat opossums with particular cruelty, and they are regular patients at City Wildlife. The previous spring the hospital received a mother who'd been pepper-sprayed. Several months prior someone found one doused with bleach. Goldberg recalled a minister asking whether they'd take an opossum family he captured in his garage. "Prior to having a place to bring them, he'd put them in a plastic bag and stone them to death," she said. "He brought them in. He was relieved. He said, 'I'm going to tell my congregation!'"

In the kitchen Peggy Hammond, the hospital's wildlife care assistant, gave breakfast to a baby squirrel. He put his whole body into the feeding, stretching out in her hand like Superman in flight to suck formula from a tube. The sight recalled a passage by Barbara Smuts, the anthropologist, about how living with baboons changed her view of animals, transmuting each from examples of a species into distinctive individuals—a realization manifested in an encounter with a squirrel. "Now, I experience every squirrel I encounter as a small, fuzzy-tailed, person-like creature," she wrote of meeting one in the woods. "Though I don't usually know this squirrel from another, I know that if I tried, I would, and that once I did, this squirrel would reveal itself as an utterly unique being." Somehow the act of feeding nourishes this sensibility. The squirrel in Hammond's palm, and all the other animals at the hospital, is no longer just one of many such creatures, but a potential someone. I don't know him well but with time I might.

Before long, a white Animal Care and Control truck rolled down the driveway. Lando McCall, ACC's wildlife specialist, entered. "We've got five wood ducks," he announced. It's a bit of an animal hospital joke, as wood ducks are notoriously sensitive and difficult to care for. I wonder why this is. The shyness that makes them so ill-suited to captivity is almost certainly an evolutionary adaptation to avoid predators, but why should predation have made them so much shyer than other ducks? Perhaps it has something to do with the exquisite plumage of males, whose emerald heads, speckled chestnut breasts, and almond flanks, traced in white with a splash of iridescent blue on each wing, made them so desirable that people hunted them to near-extinction by the early twentieth century. The species is now common again, but maybe it was only the wariest, the most easily spooked by a wrong noise or unexpected motion, who passed through the bottleneck. Maybe they're so hard to care for because of what our forebears did.

At any rate McCall didn't lift wood ducks from the kennel but rather five mallard ducklings whose mother was run over. Burbank and Hammond inspected them while Goldberg explained to another visitor, who found a fledgling blue jay in her yard, that fledglings don't need to be rescued. "She doesn't understand why someone has taken her baby. She'll be frantic," Goldberg explained, though this time she agreed to take the bird. The neighbors all had dogs in their backyards and one might eat him.

The conversation brought to mind the ecological implications of all these well-meaning interferences. Harsh though it is, fledglings and other injured animals, like the baby groundhogs I found, are an important source of food on the landscape. They're part of processes that, without humans around to interfere, support communities more bountiful and resilient than anything we could design. Taken to an extreme, what might intervention mean for local ecologies? Might urban webs of life constrict, supporting fewer predators and scavengers, and, in time, ecosystems less regulated by their biodiversity-enhancing presence? It's an interesting theoretical question. Rather less theoretical are the turtle feeding rounds, which were about to begin.

There were three in the hospital: a large snapping turtle who was hit by a car, a baby snapper inappropriately caught beside the Anacostia River and later confiscated, and a female box turtle found the previous Friday having tumbled off a ledge by the Woodrow Wilson House museum. The big snapper was in rough shape, with scrapes still raw on the thick green skin of his face and legs. Hammond held him carefully aloft, fingers safely beyond range of his eponymous bite, while Burbank injected antibiotics into a hind leg.

He seemed so out of place there, so visibly a creature of mud, slow-moving water, and slow-moving time, who by his size was at least several decades old—older perhaps than Burbank, who was 22. How did he experience that world of tile and fluorescence and hypodermic needles? People tend to underestimate the intelligence of turtles, their capacity for feeling. Perhaps this is because they're so very different from us—indeed reptiles as a whole, in the words of ethologist Gordon Burghardt, are "thought of as instinct machines"—and the frisson of danger made it easy to suspend empathy. But there was no mistaking the distress in the yawning keen of his great dignified head.

The box turtle, at least, was recovering nicely. Bandages covered the crack in her shell, which like her body was a dappling of molten orange and deep green. She almost seemed to present her arm for Burbank's injection. Just last fall, said Goldberg, the hospital cared for a male box turtle who'd suffered a similar fall at the museum. "We put him back in October. He brumated"—the turtle version of hibernation—"and is fine now," she said.

Goldberg told a story she heard of how Woodrow Wilson's wife Edith was not fond of children and, when visitors came over for tea, would send the kids outside to look for a turtle in the garden. Box turtles can live for more than a century. "The male we took care of last year might have crossed paths with President Wilson," she says. "It could have been the same turtle."

◆　◆　◆

LUNCHTIME BROUGHT A LULL. OUTSIDE IN THE MIDDAY heat, a catbird taking his daily constitutional clearly felt ready to leave. Houdini the squirrel took a nap. The ducklings, calmed by their swim, clustered contentedly in one corner of their tent. As Hammond and the interns chopped veggies and mixed formula for afternoon meals, a call came in: another turtle, found inside an auto shop garage near Rock Creek Park.

From the description, Goldberg couldn't tell if it was a painted turtle or a red-eared slider. It's a distinction with more than taxonomic significance. Per their contract with the Washington, DC, Department of Energy and Environment, which provides about one-third percent of City Wildlife's $588,000 budget, the hospital isn't supposed to take sliders. They're not historically native to the region and are considered invasive.

It's an ongoing tension: if each individual life matters, if empathy and compassion should guide one's response to the suffering of another being, why should an animal be disqualified from care by historical circumstances beyond their control? It's not a red-eared slider's fault that, following the Teenage Mutant Ninja Turtle craze of the early 1990s, they became fashionable and soon discarded purchases. One might also argue that they've become a part of their local ecologies—naturalized citizens, if you will—and that their invasive label, at least in some places, is more about our sensibilities than any documented displacement of historically native species. Inasmuch as intelligence makes ethical regard more urgent, consider this: one recent study of red-eared slider vocalizations revealed no fewer than 12 distinct call types used in underwater communication. There is a richness to their lives that science has only started to investigate.

On that day, at least, all this was academic. A photo came by email: a box turtle. "The girls are roaming now, looking for a place to lay eggs," Goldberg said, and left a message asking the turtle's finder to call back so she could instruct him on where to release her. Turtle homing instincts are exceptionally powerful. If not put back in her neighborhood, she'd still try to return, crossing roads to get there.

I wandered over to the examination room and found Burbank using a plastic tube to feed a baby rabbit the size of a peach pit. It took a while; the meal amounted to more than a tenth of his body weight, in keeping with the feeding habits of rabbit mothers, who return only a few times each day to the grass nests where they hide their babies. Burbank's movements were delicate and she spoke quietly. "They're so high-stress," she said. "Even adult bunnies can die from a heart attack if you do something wrong."* Meanwhile Goldberg was back on the phone, talking to a woman who had put a fledgling starling on a diet of chopped-up ham, apparently because

---

* The day was barely half over and Burbank, who planned to go to veterinary school in the next few years, had already cared for three species of bird, two species of turtle, and two mammal species. Veterinarians sometimes joke about how easy doctors have it, since they only have to know one species; wildlife rehabbers tease veterinarians for only knowing several.

it looked like worms. With kindly patience practiced in those years as a physician's assistant, coaxing stubborn patients to follow doctor's orders, Goldberg bottled her exasperation until hanging up. "It's not coming from a bad place," she said. "It's just misguided."

In the nursery room, a row of fledglings in incubators—five robins, a starling, and three house sparrows—received proper meals from a volunteer named Lori Smith. A biologist at the Smithsonian National Zoo, Smith until a few years ago helped care for the zoo's birds. Now she worked in the commissary, overseeing the preparation of their diets, and fulfilled her yen for providing hands-on care at City Wildlife. The fledglings had faces that, as they say, only a mother could love, with oversized, beady-eyed heads and half-formed feathers. Smith murmured maternally as she tweezed worms into their outstretched mouths. "I can do baby birds all day long," she said, moving slowly down the row and then starting anew.

Smith also volunteered each year to count red knots at their migratory stopover in nearby Chesapeake Bay. It's hard to think of a more striking contrast than between red knots—elegant shorebirds who each spring and fall fly 9,000 miles between the high Arctic and South America, their collapsing populations on a trajectory to extinction—and scruffy, ubiquitous house sparrows and starlings. I asked whether these missions ever seem incongruous. "When they get in trouble, it's on us to help," Smith said simply.

Many wildlife hospitals won't accept sparrows and starlings, who like red-eared sliders are naturalized but considered invasive. City Wildlife takes them all, using private rather than District funding for their care—a procedural shuffle around an ethical dilemma. Certain other species remain verboten, though: deer, who require facilities City Wildlife lacks; Canada geese, whose predilection for aquatic vegetation has hindered the city's wetlands restoration efforts; and species legally designated as pests, such as rats and mice. I find this unjust, but Goldberg is diplomatic. She understands why some people object to spending money on such common animals. Wildlife hospitals have been criticized for diverting funds from more important causes. "Single animals get care," wrote journalist Emily Sohn in *Aeon*, "money that could be spent instead on protecting habitats and other conservation efforts."

Sohn's critique, which is sharp and well-researched, described wildlife rehabilitation as "a feel-good industry that has evaded an objective audit for years." It's absolutely true that when scientists track the fate of rehabilitated animals—something City Wildlife doesn't generally do—the survival rates are often low. Sometimes they're simply too damaged ever to recover, as seems to be the case with seabirds rescued from oil spills, of whom one study found that barely 10 percent survived for more than a month after being cleaned and released. Young animals who spend formative months in captivity miss out on important learning experiences; had my baby groundhogs survived, they wouldn't have learned from their mother how to avoid predators and find the most nutritious plants. They might have ended up as a heartwarming Instagram post still accumulating likes long after they'd been run over or eaten or entered hibernation without enough fat to survive winter.

Sohn quoted Craig Harrison, author of *Seabirds of Hawaii*, lamenting how the rehabilitation of a single bird may cost $200,000. "If you could save an entire forest or a small ecosystem with hundreds of creatures, wouldn't that be a better use of money than figuring out how to fix a bird with a broken wing so that it can fly again?" asked Harrison. Though City Wildlife doesn't spend so extravagantly—they care for about 1,800 animals yearly, which puts per-animal spending around $300, not including donations of such items as food and bedding—the larger point resonates. The population of Washington, DC, is expected to swell by 40 percent during the next several decades; a lot of habitat is threatened. It won't be saved by caring for sparrows at City Wildlife.

Even so, this conflict between helping a few animals or saving ecosystems doesn't seem so clear-cut. It reminds me of discussions about endangered species triage, and how we must select those we value most because it's not possible to save them all. That may be the most pragmatic decision, but it also reifies as immutable truth the choices societies make to consign species to extinction. It *is* possible to save them all, but it's not a high priority. And is it really a zero-sum game, in which any money devoted to rehabilitation is subtracted from landscape-scale conservation? The idea of people canceling their wildlife hospital donations and putting that money into local watershed preservation strikes me as wishful. If anything, it seems more likely that they would support both and that compassion for individuals leads to a broader, ecological ethic of care.

Truly successful conservation requires broad public support—and in a city, acquaintance with nature often begins with common animals of little conservation value. First sparrows and starlings, then bitterns and the importance of wetland pollution regulation. "Appealing to that sense of compassion could do nothing but expand interest in conservation," says Tim Beatley, an urban sustainability expert at the University of Virginia and founder of the Biophilic Cities movement. This idea is known as the "pigeon paradox," a phrase coined by conservation biologists in the mid-2000s.

Asked about the pigeon paradox, Erin Luther was quick to say that care for those creatures should not be regarded merely as a means to some more important end. They are themselves important. That said, care for them is a powerful force indeed, and Luther recounted a story from her time at the Toronto Wildlife Centre. A woman called after finding an injured bird on her way to work. "She said, 'I was just going to pass by, but then it looked at me. We made eye contact. And I just knew that I needed to do something,'" Luther recounted. The first step was to figure out the species, and Luther asked the caller for a description: slate gray, with a beautiful iridescent purple-green sheen around his neck. "Is it a pigeon?" Luther asked. "Oh no," came the response. "This is a very special bird." He was smaller and more delicate than a pigeon; perhaps he was some sort of rare species.

"I said, 'Okay, why don't you bring it in to the hospital and we'll take a look,'" Luther continued. "She brought it in, and I took the box, and I opened up the box—and sure enough, it was a pigeon. And it was the most standard-looking pigeon you could imagine." I laughed, but Luther went on. "I was so moved. There was something about this moment of looking that made her see this very ordinary bird as extraordinary and beautiful," she said. "I always liked that story as a kind of metaphor for what happens when we pay attention to the life around us and what possibilities are there."

To Luther, wildlife hospitals are part of an emerging infrastructure that serves and shapes relations to nature in cities. Through that lens, the conversations Goldberg had throughout the day, delivering cheerful, patient advice about fledglings and turtle release and keeping cats inside, are lessons: about natural history and more subtly about coexistence, about animals as beings to be understood in terms of species and ecology but also as persons, gently nourishing an attentiveness to nonhuman life. City Wildlife responds to about 3,000 such requests annually. One can imagine each interaction rippling outward, affecting everyday choices, conversations, and social norms, in aggregate amounting to a cultural shift that might someday yield a city—a world—that, in Beatley's words, is both humane and rich in nature.

Goldberg recounted the counsel of a wildlife rehabber friend. "When you deal with people bringing in animals, she calls it your five minutes to change the world," she said. "When someone brings an animal in, they're connected."

◆　◆　◆

CITY WILDLIFE ALSO COORDINATES LIGHTS OUT DC, part of a nationwide movement to turn off unnecessary nighttime building lights that disorient migrating birds. Anne Lewis and Jim

Monsma, the hospital's cofounders, started the District of Columbia's program in 2010. One spring morning I accompanied Lewis on a shorter version of the four-mile walks she and other volunteers make each day at dawn during migration seasons, looking for birds who have finished their night's journey by flying head-first into downtown windows and falling to sidewalks below.

Altogether they've found more than 5,200 birds. That's only a small fraction of the true number of victims, most of whom hit unmonitored buildings or fly a short distance and die elsewhere or are swept up by sanitation workers or scavenged by crows before anyone counts them. Even so, it's still a useful sampling of a problem that kills an estimated 600 million birds each year in the United States. Sometimes City Wildlife is able to treat recovered birds. Often they're already dead or beyond help, but counts can still be used to identify buildings that are especially dangerous.

Through Lewis's eyes the built environment was a landscape of avian risk. At one office building, officially designated as LEED Gold for its environmentally-friendly construction, an illuminated atrium conspired with glowing red exit signs—the color of flowers favored by hummingbirds—to lure those tiny miracles to a midair end. (That building's maintenance crew alerted Lewis to the problem; custodians and other early morning workers are often quite helpful, she said.) At another site, a glass walkway that spanned the street between buildings might as well have been designed to intercept unwitting flyers.

As the sky brightened it was almost perfectly reflected by a prize-winning modernist edifice whose architect evidently didn't grasp that large numbers of birds would fly head first into giant mirrors. This oversight would have been understandable a few decades ago, before dangers posed by windows were common knowledge. How is it possible now? "It's the most frustrating thing in the world," said Lewis, who is herself a retired architect.

Many of her colleagues are sympathetic, she said, but they're conservative by temperament. Although using bird-friendly windows is easy and relatively inexpensive—they can be etched with subtle patterns, for example, or fitted with ultraviolet light–reflecting film—they're not yet common. "It's probably going to take legislation to do it, because it's just not happening," said Lewis. In 2023, the city passed a law, lobbied for with tabulations from years of predawn walks by scores of Lights Out volunteers, that new commercial and large-scale construction in Washington use bird-friendly materials.

All those acts of individual care had added up to larger change. City Wildlife not only helped the animals brought to their door but went out into the city and called attention to a problem. "I didn't think of myself as an activist when I got into this," Lewis told me. "I just wanted to fix up a few birds." Yet from that impulse came a change that could save the lives of thousands.

After a few more thankfully bird-free building checks, Lewis took me to the National Geographic headquarters. There a Canada goose couple had nested on a vegetated third-story ledge. Their proud heads poked above the parapet. Lewis and April Linton, founder of City Wildlife's Duck Watch, a network of volunteers who monitor mallards who nest in unsafe places and might later need assistance, were helping staffers keep an eye on them. The eggs hatched a day earlier; the plan was to wait until the goslings jumped to ground, then escort the family to a pond in nearby Constitution Gardens.

At the moment nothing was happening, so Linton—a sociologist at the US Department of State, who unlike the retired Lewis was taking a day off work—took me on a tour of nearby mallard nests. She showed me three: one in a patch of flowers beneath a streetside tree, another in a concrete tree box, the last in the bushes of a small patio set below sidewalk level. Except for the latter, pedestrian traffic streamed around them, and a few feet away a crush of cars. Despite their seeming precarity, explained Linton, those streetside spots were desirable. Not as good as a nest by the pond, or along the Potomac River, but still mostly free of predators, and nobody bothered them. I stopped to photograph the mother in the tree box. Her nest was unexpectedly delicate, lined with down pulled from her own chest, leaving a patch of skin pressed to the eggs to better warm them. She kept a deep brown eye on me but did not otherwise stir.

That this quiet, epic act of maternal devotion could transpire without bother in such a busy place felt remarkable, the sort of thing that could restore one's faith in common human decency. An ambulance rolled past, sirens wailing. Mom didn't react. Eventually she and the other moms would lead their broods through the streets to water, whereupon their mates might find them. (Mallard fathers have a reputation for leaving moms to raise ducklings alone, but this isn't always the case, said Linton, and indeed urban drakes seemed more likely than their country counterparts to help.) If this short migration happened on a Duck Watch volunteer's watch, they would provide an escort, stopping traffic along the way.

I asked Linton where she felt her duty to intervene ended. She would not try to stop predation—"The crows need to eat, too," she said—but she would do her best to keep ducklings from being run over or drowning in fountains with sides too high for the still flightless youngsters to climb out. It occurred to me that this well-meaning assistance might enable maladaptive nesting habits, and perhaps letting those ducks perish would be a better long-term solution—but other ducks would likely take their place, and the problem would continue. That also seems unnecessarily hard hearted. If people want to help, why not welcome it? What might happen if care groups formed around many species, and networks of people particularly sympathetic to urban turtles, owls, opossums, butterflies, or whoever kept an organized eye on them and eliminated human-made hazards in their shared environment? It's worth imagining.

Back at National Geographic we were joined by Goldberg. Plans had changed; the parents had been exploring the ledge, which circled the building, and it seemed they might jump onto concrete rather than the landscaped courtyard's forgiving mulch. Instead the City Wildlife team would try to catch the goslings and coax their parents into the building and back outside at ground level. Excitement rippled through the office as word spread. Goldberg and Lewis ventured onto the terrace, equipped with umbrellas to shield off attack. "We're in broken-nose territory here," warned Lewis. "I don't really like my nose anyway," quipped Goldberg. Their worries did not materialize; it was almost, Goldberg later said, as if the geese understood that they meant no harm. They herded the goslings into a kennel and then Lewis, holding the kennel low, walked backward into the building, the parents following with cavalcade of honks.

Together they walked down a short hallway—there the geese paused to defecate—and through a broad open room of desks and cubicles, the path lined by hushed staffers with phones outstretched, past a bank of photocopiers, and into the elevator. The geese followed Lewis inside, waited for the doors to close, and then exited on the first floor, where they padded under a set of turnstiles and out into the courtyard. There the family was reunited in the fountain while National Geographic staffers held an impromptu party. Some of the festivity no doubt derived from the respite the spectacle provided from workday mundanity, but there was genuine affection for the geese, too. The office had a Slack channel devoted to them; there had also been a live webcam watched by thousands of people around the world, and while Linton and Lewis had watched the nest, dozens of people stopped to chat. Would all this amount to no more than a self-satisfied pat on the back and detract from serious conservation? Or could it offer a path to deeper, broader care? My money's on the latter.

We let the geese rest for the evening and returned the next morning to walk them, Duck Watch-style, to Constitution Gardens. Lewis helped Goldberg kennel the goslings again and then handed me a high-visibility vest and a long-handled net. I would bring up the rear, helping keep the parents from straying, while Goldberg ran point on traffic and Lewis carried the kennel.

Our strange parade proceeded down 16th Street, from the leafy environs of National Geographic to a busier section lined by stores and upscale hotels, and past St. John's Episcopal Church. We crossed Lafayette Park, right across Pennsylvania Avenue from the White House, and headed down 17th Street, the White House grounds to our left and Secret Service agents watching impassively from behind an iron fence.

I intended to take notes and photographs so as to better reconstruct the scene, but that was impossible. Lewis barked at me to keep my eyes on the geese. They became the center of my attention, and only peripherally did I register the alternately bemused and delighted faces of the people we passed. The parents grew tired; the flap flap of their feet on hot pavement grew labored, and their mouths hung open in what I understood to be exhaustion. At one point a man on a children's bicycle approached. He gave the impression of being unhoused and, after asking what we were doing, pedaled off with a snort.

It crossed my mind that this was, in a society with so many pressing human problems, the height of frivolity. And yet—if Goldberg, Lewis, and Linton spent their time watching television instead of helping birds, would anyone remark? It's tempting to criticize acts of extravagant care, perhaps because they make us uncomfortably aware of how much more we could do, instead of celebrating them.

As we approached Constitution Gardens, Lewis enlisted the services of a woman who happened to be walking on her lunch break. She helped block Constitution Avenue's eight lanes of traffic and tourists gave an impromptu cheer as we crossed. Mom limped noticeably, her muscles pushed to their limit by what was almost certainly the longest walk she ever made. The cool green grass of Constitution Gardens must have been a balm for their feet. It seemed to reinvigorate the parents. They honked, the goslings cried, and we crossed the lawn and headed down to the water's edge—where City Wildlife had helped erect several ramps that waterfowl can use to climb the pond's perpendicular stone banks. Lewis freed the goslings; mom and dad rushed to them. Together they paddled away.

◆ ◆ ◆

IN A REFRIGERATED ROOM IN THE HOSPITAL'S REAR was a blue tailgate-size cooler labeled, *the ones we couldn't save.* It's where the bodies went. "A freezer full of lost souls," Goldberg called them. The cooler was nearly full. On top were the morning's possums, a squirrel, a young rabbit, a pileated woodpecker, and numerous baby birds who didn't yet have feathers. They looked peaceful, and also alien—not unpleasant, but foreign to a human-engineered world of windows, asphalt, and poison. A world built around and over their own, by people who can design a smartphone or a supply chain for Himalayan salt but not an indigo bunting.

Some animals have managed, certainly. Yet even for them, death often comes from us. Most of City Wildlife's patients arrive as a consequence of human activities; they've been hit by cars or run over with lawn mowers, mauled by pets, or sickened by pesticides. In the space of a few recent weeks their Facebook page featured a duck found with fishing line wrapped around her beak, baby rabbits nearly eaten by an off-leash dog, and a pigeon who needed a toe amputated after it became tangled in string and the flesh turned necrotic. Each of these incidents was avoidable. In attending to their victims, City Wildlife is performing a sort of audit, pointing to how the city and its residents could be more animal friendly. "We've become more of a working-in-the-community organization than we ever thought we would," said Anne Lewis. "There is so much more work to do in preventing these injuries than there is in saving the animals." They're making it hard to look away, to avoid responsibility.

We have a moral obligation to help the creatures we hurt. This is why the argument that money spent on wildlife rehab would better be spent protecting habitat seems unfair, much as it would be unfair to deny medical care to a pedestrian struck at a dangerous intersection because the cost of surgery might better be spent lobbying for road safety regulations. The importance of systemic, macro-scale change doesn't obviate the importance of individuals. And, in the long run, is a society that doesn't see a lone mallard as deserving of care really going to protect much biodiversity, either?

Back at the nursery, Smith continued feeding the babies. Burbank tested whether a catbird was ready to leave. It was too soon: he still couldn't fly higher than where she released him. She and Goldberg reviewed the records of a young opossum who was recovering nicely. The phone rang again: a red-bellied turtle found on a picnic table by the Anacostia, wrapped in fishing line with a hook in his mouth. "If it's not settled way down in the gut, we can anesthetize him and take it out," Goldberg reassured the caller. I thought back to when, while fishing as a teenager, I hooked a turtle and the line broke. It never occurred to me to wonder about his fate.

The day's last patient was a mallard hatchling delivered in a cardboard box by a young woman who found him that morning by the Jefferson Memorial. "I looked everywhere for the mom," she said. She kept him by her side at work, sneaking him water and moist bread. Already she had named him Francisco. "You saved his life," praised Goldberg, who wrapped Francisco in a blanket, put him in an incubator, and told the woman about Duck Watch.

Later in the summer I read on City Wildlife's Facebook page about the red-bellied turtle's release, which was attended by a group of visiting schoolchildren. Scientists who study turtles say they have remarkable long-term memories; some of this research, in fact, has involved red-bellied turtles. I like to think he'll remember them well.

9

Living with Coyotes

ONE APRIL DAY IN 2015, IN A BRAMBLE-HIDDEN HILL-side den in a park in San Francisco, a coyote named Scout was born. Protected by vegetation and topography, the hillside offered seclusion, even amid the bustle of America's second most densely populated city. There Scout's parents left her while they hunted and, after about a month, when Scout's eyes had opened and she could exit the den on her own, it made for an ideal nursery.

She spent hours by herself—coyote litters typically contain multiple pups, but Scout was an only child—playing with sticks and chasing butterflies. She explored thickets of willow and wild cherry, walked on logs, splashed in a nearby brook. Sometimes Scout just sat and watched.

It would have been idyllic but for her father. He was domineering, even harsh, protective of his mate and daughter but rarely affectionate. He was quick to assert his primacy, sometimes demanding displays of subservience. Where other coyote families spent much of each day together, grooming one another and playing, in this family those moments were brief. By the time she was six months old, still just a juvenile in coyote terms, Scout had learned to avoid her father, who had started biting her stomach and haunches—not hard enough to injure but still enough to hurt, seemingly for no other reason than to remind her who was boss. Afterward Scout's mother consoled her, but she too was forced into obeisance. She was just two years old herself, a first-time mother, maybe a little overwhelmed. In January 2016 Scout decided to leave.

The odds against her were long. Coyotes usually disperse as yearlings after spending their first winter and the following spring and summer with their parents. Even with that extra competence most of them die, especially in cities where territories are few and speeding automobiles are many. Yet fate smiled: a nearby territory, featuring a small park of its own, was vacant. Scout moved right in.

For two years she lived by herself. It was a lonely but magical time. She liked to start her days—or end her nights—atop a large hill, running in circles to greet the sun. She was smart and quick; after finding the day's food she had plenty of time left to entertain herself. She invented games, such as curling into a ball and rolling down a hillside, or stalking sticks as though they were prey. She tossed dirt clods in the air and chewed on plastic bottles, which made the most delightful sounds. She chased bees. Once she found a bicycle tire: now *that* was a treat!

Every so often Scout jumped straight up and embarked on a round of frenetic back-and-forth sprints, or what people call "zoomies" in our own canine companions. She ran after crows and ravens when they landed on the ground—not to hunt, but as play, though they enjoyed it less than she. What she really needed was another coyote. Accustomed as she was to solitude, coyotes are still deeply social creatures. If she caught a human smiling and watching her antics on the hill, she would redouble her efforts and then glance over, as though glad of the audience. There were dogs, too, walked by their humans, and she liked to watch them. Some she came to know, such as a pair of Yorkshire terriers who barked ferociously whenever they saw her, and into whose yard she snuck at night to drink from an ornamental pond and eat fallen persimmons.

Off-leash dogs sometimes chased Scout, but she was too quick and nimble to be caught. A few she liked—not becoming friends, exactly, but trotting nearby during their walks—but one in particular, a big white-haired fellow twice her size, she detested. Once he slipped his leash and chased her, not playfully but with real aggression. When he couldn't catch her, he peed all over her territorial markings while staring right at her. After that she would approach the dog and his companion whenever they visited her hill, which was often, and approach them with hackles raised to scream her displeasure. She was small but she was brave, and this was her home. Fortunately the dog's companion understood what was happening. Rather than calling the city to report a dangerous coyote or taking to social media to tell stories of their lucky escape, he would keep his dog close and walk away.

Of course Scout had no way of knowing how he helped. Indeed quite a few people were looking out for her. Once a park visitor overheard some kids talking about shooting her with a pellet gun; the visitor took their picture and warned the kids that if anything happened to the coyote, she'd send it to the police. Scout's protectors also tried to stop people from feeding her. That was happening all too frequently, with some people even tossing food to her from their cars, which in turn taught her to chase them. The feeders meant well, but it was hardly necessary. There were plenty of gophers to eat, and lizards and mice, and once an opossum. Depending on the season there were seeds, nuts, and berries, not to mention fruits from backyard trees: pears, apples, loquats, persimmons, sapote, and figs, all lovingly watered even in the midst of a drought. The neighborhood was a cornucopia.

So it went for nearly three lonely years. And then, one early September day in 2018, along came another coyote, a younger male born the previous year in a territory five miles away and driven out by his brothers. For the first time Scout had a companion. She fell head-over-heels in love: an apt description, given how much they wrestled. If some people might feel uncomfortable attributing that state of being to coyote, well, they should have seen these two.

Scout and the newcomer—call him Hunter—spent nearly every waking moment together. They trotted through the territory side by side, staring into one another's eyes, species-transcending grins on their faces. Sometimes they became separated for a few minutes while hunting and they rejoined with a burst of affection, making up for the lost time with extra muzzle-rubbing and bouts of wrestling.

In November Hunter broke an ankle. Scout took him to a safe spot, out of the way of foot traffic, and stood guard against any dog who might come near. It took a month to fully heal, during which time the coyotes' human friends intervened when well-meaning neighbors contacted a wildlife rehabilitator about trapping Hunter and trying to treat him. That would have been a disaster: intensely stressful for him and also for her, with her newfound companion taken away with no explanation. No longer was she alone.

◆ ◆ ◆

WE KNOW SCOUT'S STORY BECAUSE OF JANET KES-sler, a self-trained naturalist who, while walking with her dog in 2007, came eye to eye with a coyote. She had never seen one before; *Canis latrans* had been extirpated from the San Francisco Penin-sula in the early twentieth century and only recently returned. Kessler was transfixed. The coyote's gaze was so intelligent. So *present*. Observing coyotes soon became a passion for Kessler, a part-time legal assistant and full-time mom whose sons were now grown. In the years since, Kessler has documented San Francis-co's coyotes with extraordinary rigor.

Hardly a day has passed without Kessler spending hours watching coyotes from afar: recording them with telephoto lenses and trail cameras, studying their behaviors and interactions, tracing their relationships and life histories, mapping their territories, and telling their stories on her blog. The closest analogue is the work of Stan Gehrt, a biologist who in 2000 started a long-term study of coyotes in metropolitan Chicago, and whose name is, in wildlife circles, synonymous with the study of urban coyotes.

It's a relatively new field of research, just as urban coyotes are a relatively new phenomenon. Until the early eighteenth century, coyotes lived primarily in the prairies and deserts of central North America and Mexico. With the extirpation of wolves, they proliferated across much of the continent and, in the last few decades, they have become common in cities and suburbs. There they have challenged a common stereotype, most often heard in arguments about wolves and grizzly bears, that city people love predators in principle but don't actually know what it's like to live with them. Urban coyotes turn that notion on its head.

Granted, they're not as potentially dangerous as grizzly bears or wolves—although it should be noted that wild wolf attacks are exceptionally rare, with only two recorded fatalities in North America since 1893—but coyotes can still be scary. They do sometimes eat pets. Once in a great while they'll bite someone, though usually these are warning nips misunderstood as an attack. They add a small but real element of risk to daily life, and accepting their presence requires more effort than adopting humane approaches to managing so-called pests or supporting the treatment of injured animals in wildlife hospitals.

Shelley Alexander, a conservation biologist and wild canid specialist at the University of Calgary, calls coyotes a "sentinel species" for nature in a world where there's no longer a neat divide between human and wild systems. Their urban fate is shaped by human attitudes: about what and who belongs where we live, what constitutes a trespass, and what degree of conceivably avoidable risk or inconvenience we're willing to tolerate. They embody a principle that applies to a great many animals: their survival is as much about sociology as it is about biology.

Kessler works at that nexus. Whereas scientific studies of urban coyotes typically rely on using radio collars to track the canids from afar, Kessler depends on visual observations. She has learned to identify coyotes by sight, coming to know them as individuals with distinctive personalities and rich personal histories. She also gives them names. "I don't see coyotes," she told me. "I see Scout. Peter. Scooter." Her methods create a certain intimacy: not in the sense of interacting with coyotes, which she emphatically does not, but in the type of knowledge they yield.

Her observations overlap with scientific investigations into questions of biology and ecology—what coyotes eat, how they move about the landscape—but Kessler is most interested in the question of what it is like to be a coyote. Not just what they do, but who they are. She likens them to hobbits, a slightly clumsy analogy intended not to illustrate their character—which is certainly more energetic than that of Tolkien's easygoing halflings—but how their society takes place amid our own, a separate but parallel world.

"Coyotes ultimately are not so different from us," Kessler says. That consideration shapes her advocacy. She has spent nearly as much time talking to people about coyotes as she has watching them, all with the aim of creating what might be called a culture of informed coexistence between humans and coyotes: one that doesn't rely on human obliviousness and coyote resilience but that involves people accepting their presence and understanding enough about them to live peacefully together. When she talks to people, she provides two pieces of information: instructions on how to handle coyote encounters and accounts of coyote family life. "That way you can create a bridge," she told me. "You can relate to them."

That people in a dog-loving society should need help relating to coyotes—who are, after all, close cousins of our fur babies, and in the eastern United States often possess domestic dog DNA inherited from past interbreeding—speaks volumes about how wild canids were long regarded by Western Europeans and their colonial descendants. Wolves offer a particularly instructive example. In the New World, they were not merely killed to protect livestock or obtain fur, as were many predators, but as Barry Lopez wrote in *Of Wolves and Men*, "with far less restraint and far more perversity. A lot of people didn't just kill wolves; they tortured them." In the process of driving wolves to near-extinction, settlers set wolves on fire, dragged them behind horses, fed them fishhooks, captured them, and then released them with mouths and penises wired shut. They severed the tendons in wolves' legs and then watched gleefully as dogs tore them to pieces.

By the late twentieth century there were relatively few wolves left to kill. Echoes of that animus remained, however, with coyotes. Though not subject to the same degree of cruelty as wolves, they were—and often still are—treated as though their lives have no value. By one widely cited estimate, 500,000 coyotes are killed each year in the United States: roughly 60,000 by the federal government's Department of Wildlife Services on behalf of farmers and ranchers, the rest by hunters, trappers, and people who simply want to kill coyotes. Most states allow people to kill coyotes year round, day or night, with few limits on the methods used: electronic calls that mimic their voices, baits and spotlights, semiautomatic rifles equipped with night-vision scopes. Ethics of fairness that guide other forms of hunting do not extend to them.

Sometimes they're killed in contests that end with their bodies piled on flatbed trailers. In some places they're captured and then released inside arenas for dogs to maul, chased down on snowmobiles, or shot from helicopters. Arguably these practices receive attention disproportionate to the number of coyotes actually killed in them. More typical attitudes are described by Shelley Alexander, who surveyed landowners in an agricultural region of southern Alberta. She found a range of views, including people who killed out of a sense of necessity, fearing for their pets or livestock, or to keep what they understood as balance, neither remorseful nor glad of it.

Yet Alexander found that some people delighted in killing. They boasted of poisoning coyote pups while their mother was away, holding killing contests, using coyotes for target practice, trapping them, and cutting off their paws. Some said that killing is something one does "to feel right" and that "coyotes are one thing that you can kill whenever you want." That such practices exist and are generally tolerated by wildlife management agencies—even when recognized as objectionable, they don't threaten aggregate populations—speaks to how little value coyote lives usually have.

While it would be wrong to characterize these as intrinsically rural practices—for many country dwellers respect coyotes and some city people loathe them—it is fair to say that people's worst impulses are more easily indulged within rural management regimes. Regularly trying to shoot, trap, or poison animals in densely populated areas simply poses unacceptable risks to humans and their pets. Moreover, a different ethos prevails in cities. Urbanites tend to see wild animals in a more neighborly light and municipal governments, rather than state agencies established to regulate hunting, handle the day-to-day job of wildlife management.

This arrangement has created an opening for coyote advocates. Groups such as Project Coyote and the Humane Society, and scientists such as Gehrt, who was mentioned earlier, advise cities and citizens on how to coexist. Conflicts do happen, and coyotes may be killed for threats real or perceived, but in many places acceptance is the baseline. Coyotes appear in the Instagram feeds of wildlife hospitals and heartwarming viral videos. The routine persecution seen elsewhere is conspicuously absent.

Nonetheless it would be misleading to think that this harmony was fated, or that it's guaranteed into the future. It's an ongoing project that requires education, communication, and public competency. At one point in our conversations, I became confused by

Kessler's frequent references to coyote management; I associated that term with setting population targets and using lethal methods to attain them. "Oh no," she said. "A coyote management program is how to get along with them."

◆   ◆   ◆

SCOUT AND HUNTER SPENT FIVE BLISSFUL MONTHS together. Then, in the cold and damp of the San Francisco winter, another coyote arrived on the scene. She was one of the few whose origins Kessler did not know; the newcomer had first appeared in the territory of a pair named Cai and Yote, and a month earlier she'd wandered into the Presidio, a former army base turned national park in the city's northern tip. There she was captured by researchers and outfitted with a GPS-enabled radio collar that allowed them to track her movements.

Kessler called her Wired. She learned of her arrival in Scout's territory after seeing Scout running pell-mell down the middle of a street, as if in pursuit, and then, later in the day, struggling to climb her hill. Through a camera lens Kessler saw gashes on her face. A neighbor told Kessler about seeing two coyotes fighting. Blood was smeared all over the sidewalk and underneath a car where it happened. Over the next several weeks the coyotes fought again and again, with Scout taking the worst of it. Neck bites are a basic coyote fighting tactic and, with Wired's collar functioning like armor, Scout never had a chance. In a month's time she was a skinny, mangled, dispirited mess, her fur matted by the seeping of infected wounds.

Finally Scout could fight no more. For a time she rested in a nearby neighborhood where one of Kessler's friends, a teacher of Buddhism named Cynthia Kear, kept tabs on her. ("In Buddhism, we have this model called a Boddhisattva: someone who lives a life of service and wants to help all beings," Kear told me. "Janet revealed herself to me as a coyote Boddhisattva.") Kear talked to neighbors and made sure they left the coyote alone. It brought the community together, said Kear, and from her desk window she could sometimes see Scout lying on a strip of grass across the street. When I talked to Kear, years after this happened, the memory was strong enough to bring tears to her eyes.

Eventually Scout regained her strength. Now she was without a territory, scraping out a living on the margins of others. Kessler followed her sojourns with camera traps set up where people told her they'd caught a glimpse. Sometimes Scout gathered enough strength to make a foray into her old home, only to be beaten back by Wired, and sometimes Wired ventured out and tracked her down, even miles away, as if hunting her. Wired seemed to understand that her overmatched adversary would not quit until she was dead.

The end seemed imminent. Without a territory, even a healthy coyote's life is usually short. Scout was battered and covered with sores. When it wasn't Wired chasing her, it was the resident coyotes of the territories into which she trespassed. Finally there came an October day when Scout returned to her old home and found it—empty. For reasons known only to her, Wired had abandoned it. Yet the joy of return was bittersweet: Hunter had moved on, too, finding a new mate in a territory a few miles away.

Was Scout heartbroken? Resigned, or possibly indifferent? Perhaps she had long ago realized, when Hunter failed to defend her and then follow her into exile, that she was on her own. Maybe it's unfair to phrase things in this implicitly judgmental manner. Coyotes are, except in rare cases, monogamous and mate for life, but Wired's arrival forestalled what would have been their first mating season.

Whatever the case, by late winter Scout had found another mate: a young male of unknown origin whom Kessler named Scooter. Their relationship was not like the one she had with Hunter. They didn't gaze at length into one another's eyes or walk with their sides pressed together. Whereas Scout seemed besotted with Hunter, always eager to please, she was far more assertive with Scooter, occupying a dominant role. The magic of first love was not to be rekindled. Yet there was genuine affection, practical and solid if a bit short-tempered on Scout's part. And in April, in a den upslope from a highway and protected by a fence at the end of a dead-end street, Scout became a mother.

There were four pups: Peter, Flopsy, Mopsy, and one who died young. She nursed them, and then hunted for them, and when they were old enough to leave the den, she and Scooter taught them how to find food and water for themselves, how to cross streets, what to do with dogs and their two-legged animals, and how to behave. Scout was a loving, attentive mother. They shared the family joys Scout never had as a pup. Each evening they gathered, tails wagging, to play, groom, and enjoy one another's company. The following spring, she and Scooter raised another six pups. And there, as Kessler and I drove by Scout's hill and she warned me that I probably wouldn't see anyone, sat a coyote, silhouetted against a cloudless October sky.

◆　◆　◆

WE HAD STARTED THE DAY WITH A PREDAWN TRIP TO Land's End, a forested seaside park in San Francisco's northwestern corner. It's one of 20 or so territories Kessler has mapped. A Veteran's Administration hospital backs up against it, and on the paths behind the hospital she found freshly deposited scat. She bagged it for later delivery to Tali Caspi, a PhD student in the lab of Benjamin Sack, a mammal ecologist at the University of California, Davis. Caspi would analyze the scat for use in her diet studies. Another of Sack's students, Monica Serrano, uses DNA from scat gathered by Kessler to map genetic relationships between the city's coyotes. That work is ongoing but so far supports the family trees Kessler has compiled by observation.

From there we went to the Land's End golf course. Kessler, who at 70 years old remains spritely—although the word that came to mind, given her fine features and gray-streaked blonde hair, was elfin—set a fast pace. I'm reasonably fit but had to work to keep up with her. On the golf course she showed me a dugout, hidden between fairways in a tangle of poison oak and trailing blackberries, where a coyote could hide or spend the day. There were none to be seen but a beefy man in wraparound sunglasses waiting for a bus said one came by a few minutes ago.

"There's some that's got young 'uns now," he told Kessler. "Late in the evening, the mother will bring them out. They're about maybe six weeks old." He spoke in an easygoing drawl and seemed pleased to be talking about coyotes, which made it somewhat jarring when Kessler briskly replied, "No, no, no, no. They're five months old now." The man seemed irritated. "Well these here are little ones," he said. "I been out here for a long time and I know the difference."

"I don't think so," insisted Kessler. "They're bigger than you think." The man was decidedly frosty as we left, but Kessler was irritated too. "I'm sorry," she said, "but I saw her lactating in March. They can only have pups once a year and they were all born in April. But you saw his definiteness: he *knows*."

Correcting that misperception was about more than being a stickler for fact. Had there been bystanders they might have overheard and thought that coyotes can reproduce at any time of year, which feeds narratives of being "overrun by coyotes." Kessler heard that phrase often during the COVID-19 pandemic's first months, when newly homebound people saw their local coyotes for the first time. That perception lends itself to a belief that coyotes are "out of control"—a term that arises whenever coyotes make people nervous, with predictably unfortunate consequences. "They're not reproducing all the time," Kessler said. "They're not going to be invading us." When most people see a coyote, said Kessler, they really only see the outline, and they fill in the details with rumors and hearsay and bias. "Brazen," with its connotations of disrespectful boldness, is another well-worn headline mainstay: "Brazen coyote strolls Beaverton, eats squirrel." She spends a lot of time correcting misconceptions.

During the previous winter I had watched Kessler give a talk by Zoom to a neighborhood group convened after someone was unnerved by coyotes hanging out in their driveway. She spoke about coyote natural history, behavior, and day-to-day social life, accompanying the lessons with photographs and histories of that neighborhood's family. Perhaps her audience wouldn't ever be able to identify a particular coyote as mom, dad, or a pup, but they'd know that's who the coyote *could* be. They'd know the coyote had a story. In turn they might be more willing to make the effort of coexistence, beginning with Kessler's advice that, when people see a coyote, they go in another direction. If walking a dog, keep the dog close and, if possible, pick them up. Eliminate the possibility for interaction. If one can't, it's okay to ask the coyote to move by waving, stamping a foot, and perhaps tossing a small stone in the coyote's direction, taking care not to make contact—but if the coyote doesn't back down, it means something. It has nothing to do with a coyote being "aggressive," "brazen," or having "lost their fear."

Most likely there is a den nearby. Much of what is considered an attack, or at least a threat, is coyotes defending their territory during pupping season. They confront people, or more commonly their dogs, snarling and growling, perhaps darting in to deliver a firm but not injurious nip. Often they do this after sneaking up. It's a coarse but effective form of communication that other coyotes understand as an invitation to get lost. Humans, understandably enough, tend to lose the message in translation. "Please remember their aim isn't to maul your dog. It's to get your dog to leave," Kessler pled. If the coyote follows—so-called escorting behavior, easily misunderstood as stalking or chasing—don't be alarmed. Understand that they're just making sure of your departure.

Kessler's recommendations are similar to other guidelines, but they differ in their lack of emphasis on hazing. The Urban Coyote Research Project, for example, encourages people to shout or throw something toward a coyote upon encountering one. The Humane Society describes coyotes who feel safe around humans as having "lost their fear," something that should not be tolerated.

To be clear, these guidelines still preach coexistence; the difference with Kessler's recommendations is merely one of weighting. They overlap much more than they diverge, especially on the singular importance of not feeding coyotes. Yet Kessler's emphasis on interpreting behaviors misunderstood as aggression is telling, bespeaking a coexistence founded on understanding.

Shelley Alexander, who apart from her academic research also runs a coyote coexistence program on the University of Calgary's campus, supports Kessler's nuanced take. Though Kessler might not be a scientist, Alexander believes that her rigorous observations give her an understanding that even some scientists who study coyotes may not have. Alexander laments how people are quick to equate a lack of fear with aggression, as though our very presence should send a coyote fleeing. That expectation, she said, represents a shifted baseline of behavior. Only where coyotes are persecuted do they flee the sight of humans—but that's now so common it seems normal and natural to us.

"There's a fundamental flaw in this assumption of aggression," Alexander said. Yes, coyotes are wary, and if you fix your attention on one and move toward them, they ought to move away from you—but if one doesn't, that's not aggression. It just means the boundaries are temporarily different. Yet Alexander appreciates that most people don't have a presence of mind honed by years of experience with coyotes. They need simple rules, and erring on the side of what amounts to low-level harassment may make people feel safer and lead to fewer potential conflicts.

It's a lot to ask that someone frightened by a coyote think through a short decision tree and understand snarling as communication. Yet for most of human history this is exactly what people did. Kessler cites *The Ohlone Way*, an account of precolonial life in the San Francisco Bay region, in which the historian Malcolm Margolin writes of how coyotes were common and their presence entirely unremarkable. They even entered people's homes at night to procure a snack. And by historical terms coyotes are mere pups:

cultures of coexistence between Ju/'hoan people and lions in the Kalahari of southern Africa, or villagers and leopards in the western Himalayas, hint at how humans until recently shared landscapes with large predators, and they did so by understanding the predators' behavior and negotiating engagements.

That's why Kessler's approach, with its emphasis on giving coyotes space rather than scaring them off, is so compelling. It presupposes a certain level of understanding. It asks people to possess a basic knowledge of the animals around them. This competency is needed anywhere people and large animals, particularly carnivores, are going to coexist. Persecution has now relegated large carnivores to remote places; people have forgotten how to live with them, thinking they belong only where we do not. On a planet of ever-shrinking wilderness and ever-expanding human footprint, that mentality leaves very little room for them at all.

Writing in the journal *Nature Ecology & Evolution*, ecologists led by José Vicente López-Bao call for "freeing these species from wilderness, viewing them instead as normal and legitimate parts of human-dominated landscapes." The challenge, they say, "is whether modern societies are willing to tolerate and adapt." If people can't tolerate and adapt to coyotes, what chance is there for wolves, bears, or lions?

On the path around the golf course Kessler and I met two women walking their dogs. They gave an impression of well-composed affluence; they didn't fit my mental image of the kind of person who gets excited about coyotes. One of them recognized Kessler. "I'm the one who sent you that coyote scat!" she said, meaning she had emailed Kessler about seeing poop that could be collected for her research project. Their dogs—two cheerful boxers and a shy chihuahua—greeted us happily. They seemed particularly taken by Kessler, a scene that repeated itself quite a few times during my visit. Dogs just seem to like her. The chihuahua's person warned us that she'd been rolling in something. "If you're looking for coyote scat," she said cheerfully, "it's probably on her back."

The woman with the boxers mentioned an incident they had witnessed several months ago a little way up the path. Though they see coyotes regularly and never had any trouble, this time was different. "She was quite aggressive," she said, and described how the coyote faced them down and made throwing-up sounds. "It's protective," Kessler said. "They're messaging you. All you have to do is walk away." The woman didn't need the instruction, though. One of her dogs had been on a leash, she said, and she quickly called the other back, put a leash on, and left.

◆　◆　◆

FROM LAND'S END WE DROVE TO SCOUT'S HILL, BUT IT was not Scout who I saw silhouetted atop it.* Kessler said he looked like Peter—one of last year's pups—or possibly Scooter, although from our vantage it was hard to tell.

"Today we are the heirs of that distance, and we take it entirely for granted that animals are naturally secretive and afraid of our presence," wrote Malcolm Margolin of how persecution by settlers and their descendants had made coyotes flee our presence. It occurred to me that this tableau, with the coyote sharing the hillside park with a handful of people out walking their dogs or taking a morning stroll, in the heart of a metropolis—a setting that nobody would ever call wild—actually represented something close to natural.

As Kessler and I climbed the hill we saw another coyote. The first had indeed been Peter; the other looked to be Scooter. We stood on the hilltop and watched them move across the slope in search of gophers. The unhurried proximity allowed me to savor their physical beauty: the mélange of silver, cinnamon, and cream that formed their coats, the great triangular ears, the long faces, and lean mus-

---

* I can't be more specific about the location. Kessler worries about disclosing the location of individual coyotes for fear that, as has happened before, people will seek them out and feed them, or even entice them into approaching for a selfie.

cular bodies. They trotted with an effortless, intent grace that made it seem as though they were floating just above the dried golden grasses, a grace evolved in the plains and deserts of North America, where one million years ago their ancestors coursed beside mammoths and ground sloths and bison with 10-foot-wide horns, a whole menagerie of now-lost creatures whose essence still echoes faintly in a coyote's gait. As a rule I don't like to think of animals as symbolizing anything—that way lies an unfair villainy—but I could not look at them without feeling a visceral thrill of wildness.

It was midmorning and the park became steadily busier. Scooter departed but Peter remained near the path, and Kessler and I watched as walkers passed within a few feet of him. I expressed my delight at seeing humans and coyote sharing a space so intimately. Kessler, though, was not pleased. "It's horrible," she said. Coyotes and humans should not fear each other, but wariness was quite healthy. They ought to keep a greater distance than this. Such a carefree mingling—especially with so many dogs around—was bound to result in conflict, real or perceived, and the sort of fright that leads to angry social media posts and calls to the city. At one point Peter followed a man and his two dogs, a German shepherd and a smaller white terrier-type, until the shepherd chased him off. "People get used to them like that," Kessler said, and then things go wrong.

Next morning we again met at the hill. Kessler had somber news: officers from the city's Animal Care & Control department found a dead coyote by the freeway, not very far away, the body so mangled that they couldn't even identify a sex. Kessler feared the worst. She hadn't seen Scout for more than a week. This would explain why. "I've been through a number of deaths, and all of them broke my heart," she said. But Scout's death would hit especially hard. Scout was not a pet, but she had become someone special. Still, so long as the body couldn't be identified, there was a chance she was alive.

As we clambered up the hill, Kessler received a text message from Jeannette Raye, a pest management specialist who helped manage the grounds of a nearby reservoir. There was a coyote there. We hopped in her Prius and drove a short distance over. Upon entering the gate, we saw Peter. He kept his distance, trotting away on the path that ringed the water tank. The reservoir was roughly the size of a football field, surrounded by a fence that made it an island of relative seclusion amid the neighborhood bustle, and Kessler showed me a den Scout and Scooter had made this year. It was located on an embankment with a preschool on the other side. It would have been understandable, albeit unnecessary, for the school's administrators to be alarmed, but they'd merely filled in a hole under the fence through which the coyotes could wriggle. There was no reason for alarm.

That sensible response was due in no small part to Raye, who liked the coyotes and kept a protective eye out for them. A tall woman with a laugh-lined face and fisherman's beanie, Raye greeted Kessler warmly as we approached the storage-container offices where she waited. She pointed out a dugout beneath one of the offices. There would be construction soon, and she wanted to make sure that the pups were old enough to move on safely from the disturbance. Kessler reassured her that they were. Then Raye indicated where she'd seen one of them bury something that morning. Kessler put on a latex glove, got down on her knees, and started digging. A moment later she held the prize aloft: a gray-and-black striped cat's paw.

◆　◆　◆

KESSLER DOESN'T LIKE TO CALL ATTENTION TO THE fact that coyotes eat cats. It's not the sort of thing that makes them popular. As a cat person myself, I understand the grief and fear people feel about losing their feline friends. Yet it's also indisputable that when people allow their cats to roam freely outside, they themselves are predators. In the United States, cats kill an estimated 2 billion birds and 14 billion small mammals every year. To be upset at coyotes for doing precisely what cats are doing is unfair. And in areas where feral cats are common, coyotes bring ecological regulation to landscapes that would otherwise be impoverished by those delightful, voracious creatures.

Coyote ecology is a favorite subject of Jonathan Young, a young ecologist with the Presidio Trust, which stewards the Presidio's 1,500 acres of northern San Francisco verdancy. There he's responsible for all sorts of things: restoring native vegetation, reintroducing species, and generally tending to nonhuman life. He has an especially soft spot for the coyotes whom it's his job to manage. Partly they're just marvelous creatures. They also bring life to that urban haven.

He invited me on a walk that started auspiciously, with a jogger excitedly alerting us to a coyote ahead. We caught a distant glimpse of him trotting through a copse of eucalyptus, unbeknown to a woman carrying a baby on her back down a trail a few hundred feet away. Then Young led me across a grassy slope—a serpentine grassland, he said, referring to the unique vegetative community that grows from the mineral-poor soils that form over serpentine bedrock—to a sheltered spot where, upon closer inspection, coyote scat and small bones dotted the ground. He handed me a jointed series of bones that I did not recognize. "A cat's hip bones," he said.

Before coyotes returned to San Francisco, said Young, the Presidio had "a tremendous feral cat problem." Raccoon numbers were also exceptionally high. These swollen populations reflected what ecologists call a mesopredator release: the absence of larger predators results in a dramatic proliferation of small- and medium-sized predators, with corresponding impacts on prey species and the webs of life in which they're entangled.

Describing these impacts can be a somewhat thorny matter. Conservationists point to estimates that cats have contributed to the extinction of 63 vertebrate species. Cat advocates rightly note that most of those extinctions occurred on islands, with population dynamics that don't necessarily translate to mainland habitats. They also argue that it's seductively easy to scapegoat cats instead of confronting the deeper problem of habitat loss. The question remains, though, of how best to tend those habitats that are still intact, and extinction is not the only metric. There's also the matter of local extirpations and population declines, of constrained possibilities for life to flourish.

Thornier still is the question of what to do about free-ranging cats. Many conservationists call for lethal control, but this is understandably controversial and in many places there is little public appetite. It poses an ethical dilemma: Is it ever truly just to kill some animals so that others may live, simply because we have a preference for the latter's presence? Even if some amount of killing can be justified, what happens when, as would be the case with cats, it must be conducted in perpetuity, so that biodiversity is maintained only by a state of permanent warfare? Lethal control is also logistically difficult. Trying to poison cats would harm other animals, and though there are wildlife sanctuaries in the region where predators are shot and trapped to protect rare birds, those methods are ill-suited to a park as busy as the Presidio, which receives millions of visitors every year.

Coyotes offer a better solution. Young considers them apex predators, large enough that nobody else eats them, whose presence shapes the ecosystems around them. By historical standards they're small for apex predators, a term typically associated with lions, grizzly bears, or wolves. One million years ago, though, even those animals were midsized. The baselines shifted.

Studies show how coyotes reduce nearby cat populations and increase the diversity of other species, something they accomplish—the bones in Young's hand notwithstanding—less by eating cats than by discouraging their presence. Young said that coyotes now keep cat populations, and also populations of raccoons and rats, at levels that allow other species to flourish. In the Presidio, said Young, the victims of mesopredator release included California quail, an endearingly social, ground-nesting species who were unable to reproduce faster than they were consumed and vanished in the early 2000s. It might now be possible for them to be reintroduced.

Young also showed me a scat full of elongated, wrinkled seeds. Their provenance stumped him and Tali Caspi, the student studying San Francisco coyote diets, until a gardener identified them as date palms, an originally Mediterranean species now naturalized in California. For all the attention that predation receives, coyotes eat a great deal of fruit; in the Presidio this includes pears, huckleberries, cherry plums, and coffeeberries, the seeds of which they deposit with a helpful pile of fertilizer. Coyotes also make landscapes more verdant.

I followed Young down a serpentine hillside—one of the only places in the world, he said, where blooms *Presidio clarkia*, an endangered species of primrose—to one of the park's precious freshwater seeps. There in the shade of a willow thicket the Presidio's coyotes liked to rest. Often Young would find dog toys and shoes—coyotes, like dogs, seem to have a thing for shoes—that they'd found and taken here to play. We rested for a while ourselves, and conversation turned to the human side of coyote management.

On our walk he had showed me the signs warning people with dogs to stay away from certain trails during pupping season; if someone ignores the signs and has an encounter, they can't say they weren't warned. If someone reports an incident, Young will call them and conduct an interview to determine precisely what happened. "People are really emotional," he said. "What we're trying to do is get through the emotions and really assess what was happening."

Terminology is very important to Young. He doesn't use words like "attack" or "aggressive" lightly. "I can't even tell you the amount of people that call and say, 'I was attacked.' And it's like, 'That's a big deal. What happened? Are you okay? Did you have to go to the hospital?' 'Well, no, it didn't bite me.' 'What do you mean you were attacked?' 'It just looked at me and I felt uncomfortable,'" Young said in a pantomime of conversation.

Nevertheless, Young stressed the importance of taking people's concerns seriously. Even if the risks coyotes pose are mathematically minuscule—there have been two documented fatal coyote attacks in North American history, compared to roughly 25 such dog attacks per year in the United States, and the proportion of attacks that cause injury is also correspondingly small—failing to address them could turn public opinion against coyotes. It could lead to popular demands for large-scale lethal control and, failing that, to vigilante justice. And while listening to people was usually action enough, said Young, in some exceptional cases killing would be judged necessary for public safety. In a way it would also be a roundabout form of coyote safety, creating a sense of security that forestalled more killing.

When authorities did make that tragic choice—something that has happened twice since coyotes returned to San Francisco—it would at least be done after a due process requiring evidence of a true attack, and they would be certain to kill that specific coyote.* "We want to be very confident that we can go in there and get the culprit, not just the first brown coyote who happens to walk by," said Young, who is part of an informal incident review committee composed of representatives from local, state, and federal agencies. That is both fair and pragmatic. Apart from the ethics of killing animals for something they did not do, it makes little sense to kill those individuals whose behavior raises no concerns. As contrast he offered an example from southern California, where authorities responded to a coyote who had developed an appetite for cats and small dogs by simply killing coyotes until they found one whose DNA matched the culprit. It took eight or nine, he said, to find the one they wanted.

While we talked, I glimpsed a ruddy wriggling in the seep's brackish water: a rough-skinned newt, one of hundreds hatched from eggs that Young and his colleagues had transplanted here, hoping to restore the species to the Presidio, where they'd winked out of existence a half-century ago. They had also released Pacific chorus frogs, whose eponymous voices might someday fill spring nights here, and western fence lizards, one of whom I unwittingly roused from basking on a dusky rock. Electric-blue damselflies perched on cattail stems and had we lifted one of several plywood sheets that dotted the ground, we'd have found garter snakes. It was an oasis of life: the sort of place that turns bio-

---

* Although Kessler's extraordinary example made her this chapter's protagonist, San Francisco's Animal Care & Control Department deserves a special mention. The department is dedicated to peaceful coexistence with coyotes; when new hotspots of coyote activity emerge, their officers respond to residents' concerns with understanding, education, and a zero-tolerance policy for feeding. Coexistence "is possible and it's become part of everyday life," says department spokesperson Deb Campbell. "Lethal removal is a knee-jerk reaction to problems that humans cause."

diversity from a dry, academic-sounding word into a wealth of beauty and wonder.

With the exception of newts, whose skin is poisonous, coyotes would eat all these creatures, but they would also eat—or at least threaten to eat, and thereby temper the presence of—the raccoons, skunks, and especially rats who also visit this seep. Take coyotes away and there would be many fewer lizards and snakes and birds nesting in the thickets. "It's all connected," Young said. "Their presence actually promotes biodiversity."

We headed back to the parking lot, passing up a hillside wreathed in, appropriately enough, a shrub called coyote brush, densely clumped with branches covered in fluffy seedheads that resembled coyotes' tails. A flitting caught Young's eyes. Two California ringlet butterflies spiraled around each other in a mating flight. They're thumbnail-sized, not immediately striking and rarely staying still enough to examine, but magnified in a photograph they are subtly gorgeous. Creatures of foothill grassland meadows, their tawny wings matched the now-dry vegetation, but their coloration changes throughout the year and in springtime would have been green. They too have been restored to the Presidio after going locally extinct; perhaps one could draw a line between coyotes and gophers, their favorite prey, whose burrowing disturbs and nourishes the soil so as to better support the ringlets' host plants, but in too-high numbers would degrade the landscape. Coyotes even bring butterflies to life.

Along the way we were treated to a vista of grasslands, forest, and old military buildings that have since been converted to residential apartments. The Presidio is the only national park in which people live, and in a way it's emblematic of the ecological patchwork of natural and urbanized, wild and human-controlled, that prevails across early-twenty-first century Earth. Young used the phrase "novel ecosystem": an assemblage of species that has no analogue in our planet's history. Over them all, coyotes reign.

◆　◆　◆

THE NEXT MORNING KESSLER TOOK ME TO A PARK IN the territory where Hunter ended up after Wired chased Scout away. There he met Mouse, whose previous relationship had ended after a careless homeowner sealed up the space under his porch, thinking to discourage the coyotes from denning there—only their pups were still underneath and ended up starving. By the time they were rescued it was too late. Kessler believed the experience was so traumatic that it caused the male to leave. Hunter and Mouse went on to have pups of their own, and Kessler said he was a devoted mate and father.

Earlier that summer Mouse had been struck by a car. Another female had been seen in the territory, so perhaps Hunter already had a new mate. We didn't see them, and from there we visited a nearby retirement home where Kessler wanted to check the trail cameras she left in a forested glen where Hunter denned. On the path we bumped into Yvonne Renoult, a middle-aged woman with a gentle demeanor who recruits and trains nurses at the home. She's yet one more person who lets Kessler know about finding scat. Renoult mentioned that she sometimes encountered coyotes on the path. I asked Renoult if she felt afraid. She used to be, she said, until hearing Kessler speak.

Afterward Renoult typed up her notes from the talk. She gave a layperson's distillation in response to fearmongering on Nextdoor, a neighborhood social media platform—her post had more than 650 likes, she said—and wrote a guide for the retirement home's new residents and nurses. The latter lived in on-premises housing and some had armed themselves with sticks in case they met a coyote while walking home. They didn't do this anymore, said Renoult, because they were no longer afraid. When Renoult herself lived at the complex she would hear coyotes howling at night. "It used to send chills down my spine," she said. "Now it's like they're singing to each other."

As much as I came to San Francisco to see coyotes and meet Kessler, it was these encounters—with Yvonne Renoult and Jeannette Raye, Cynthia Kear and the Land's End dog-walkers, and many others—that moved me most. They were everyday people turned coyote advocates and their example would slowly spread, shaping conversations and ultimately the life of San Francisco's coyotes, long after Kessler was gone. They would help make coexistence possible: a *good* coexistence, founded on consideration and understanding, and not just the ability of coyotes to survive in spite of us. This was Kessler's legacy.

"It's got its own momentum now," Kessler said of her work. "You have more and more people who now understand coyotes. I've said to them, you've got to let your voice be heard. If an argument comes up, get in there and defend them. And people are doing that." Several times Kessler talked of quitting. She had started playing the harp again, a passion she'd abandoned after nearly severing a finger in a kitchen accident, and practicing to perform at the level she demanded of herself didn't leave enough time for her coyotes. I got the sense that she was tired, that after fifteen years of daily observations and of explaining, cajoling, and pleading, she was ready for a change. It also occurred to me that she only *thought* she would stop and would inevitably be drawn back to the animals she loves. Even then, time would catch her; but she would live on.

We paid one last visit to Scout's hill. She was nowhere to be seen, although Peter was there again. He caused no trouble, not even when the big white dog who had been his mother's nemesis passed beneath his perch. The city sprawled out before us, glittering in the morning sun; in the distance we could see the freeway where Scout's first litter was born. It was a beautiful but intimidating sight. How, in that seemingly solid mass of human development, could coyotes possibly thrive? What would happen when the time came for Peter to disperse and find his own territory? It all seemed so precarious. Yet there they were, thriving.

A few days after I left news came that the coyote by the freeway had been identified. It was Peter's brother, Flopsy. Days later one of this year's pups, whom Kessler named Captain, was found dead in a field. She suspected he was killed by rodenticides. When Kessler next saw Scout, she was listless and heavy-footed, which suggested she ingested the poison as well. But still—she was alive.

IV

IN THE WILD

10

Under New Management

In 1984, Fred Koontz got a job at the Bronx Zoo in New York City. He was a "curator of mammals"—an old-fashioned title redolent of a time when animals were displayed like exhibits in a museum, with little more concern for the quality of their mental lives than one would feel for a statue. Koontz did not belong to that tradition. He had recently finished his doctoral research on animal behavior at Smithsonian's National Zoo, a hotbed of new ideas—mainstream now, radical then—about emphasizing science and the conservation of wild animals. At the Bronx Zoo he met the veterinarian who would become his wife, and they moved into an apartment on zoo grounds.

They often took their work home: mammals who were sick and needed round-the-clock care, babies abandoned by their parents. Among the nonhuman guests in their apartment were proboscis monkeys, cotton-top tamarins, pygmy marmosets, even a gelada baboon named Jenny who loved to watch Sesame Street while resting her head on her favorite stuffed animal. There were birds, too, including Amazon parrots and a ruby-throated hummingbird who arrived injured during fall migration and stayed through the winter before being sent on her way in the spring. "She was so cognitive," he recalls now. "She recognized you and got all excited." For several nights a baby California sea lion slept in their tub.

Koontz had always been interested in animals, but living with them gave him a newfound appreciation of their cognitive and emotional lives. He didn't just care for an Egyptian fruit bat; he cared for Frodo, who fluttered about the apartment and landed on his shoulder when called.

The revelation was uncomfortable. Although the Bronx Zoo and others in the industry were evolving, and they had started to devote more resources to wild animal conservation than in decades past. The welfare of individual animals in their care was far worse than it is now. Even today there's a case to be made that, for many animals, captivity in zoos remains unpleasant or downright terrible—but there has been progress. The early 1980s were the Dark Ages in comparison. "There were many sleepless nights," recalls Koontz. He was driven by a passion to protect species, "but then I had to go walk around and see these animals in cages."

Koontz left the Bronx Zoo in 1998. He went on to work for an environmental nonprofit that mentored conservationists in developing countries, then managed a nature preserve in New York before taking a job at the Woodland Park Zoo in Seattle, Washington. There he helped oversee their field conservation projects, including collaborations with the Washington Department of Fish and Wildlife on the recovery of endangered western pond turtles and Oregon silverspot butterflies. But although he was a conservationist first and foremost, Koontz retained that sense of eye-level empathy and affection for animals.

He might be understood as a conservation-minded personification of a sea change in public values: away from a sense of mastery over wildlife and toward what social scientists measuring the shift call "mutualism values" that consider wild animals to be "fellow beings in a common social community," with intentions and emotions and minds of their own. That does not mean everyone is now an animal rights activist; far from it. But ever fewer Americans think wildlife should be managed primarily for human benefit, and ever more look upon animals as persons of a sort.

When animal advocates helped lead a campaign against wildlife killing contests in Washington, Koontz was among the citizens who testified at the Department's public hearings. His message wasn't only about the wrongness of people organizing to kill as many creatures as they could, preying on species little-protected by law—coyotes, raccoons, foxes, crows, bobcats—and offering prizes to whoever killed the most, the biggest, or the smallest. Koontz also encouraged the Department not to be afraid of change. His time at the Bronx Zoo, where he'd been on the receiving end of activists' ire, and where their pressure ultimately pushed the zoo to stop acquiring new elephants, was instructive.* Sometimes public values changed. It could be uncomfortable—but it led to progress.

The commission ultimately voted 7–2 to outlaw killing contests, making Washington the seventh state to do so, followed by Maryland in 2021 and Oregon in 2023. The contests remain legal elsewhere. Although the number of animals killed—about 60,000, according to estimates from the Humane Society of the United States—is small compared to the toll of habitat destruction or vehi-

---

* While at the Bronx Zoo, Koontz knew Happy, the elephant plaintiff in the Nonhuman Rights Project's lawsuit. At the time, he believed that animals couldn't have legal rights because they don't have responsibilities. Later he was swayed by the argument that a human with severe neurological deficits might not have responsibilities either, but a trustee can be assigned to represent that person's interests. Koontz now supports some form of rights for Happy and all sentient animals. In the case of Happy, he thinks that—given her age and personality— her interests are best protected by remaining at the Zoo. Despite this personal evaluation, though, he believes the best outcome for the lawsuit would have been for the court to appoint a panel of independent experts and ethicists to judge Happy's best interests.

cle strikes, the disrespect for life such contests embody has made them a flashpoint in a much larger issue: the practice of wildlife management by the government agencies who, in a sense, mediate relations between humans and animals across much of the United States. How should hunting, fishing, and trapping be regulated, and priorities set for agencies tasked with those jobs as well as protecting other species and nourishing biodiversity? Who ought to have a voice in deciding? And how should values of kinship and mutualism, and an ethical appreciation of animal intelligence, be represented in the process?

◆　◆　◆

THE CURRENT SYSTEM OF US WILDLIFE MANAGEMENT can be traced in part to an 1842 Supreme Court decision in a dispute over oyster harvesting in New Jersey's Raritan Bay. The court decreed that the people of New Jersey collectively owned its lakes, rivers, and intertidal coastline. Those waters, and the land beneath them, were too important to be privately controlled. Instead they would be managed by the government for the common good.

It was a landmark in what is now known as public trust doctrine, which would eventually be extended to wildlife as well. Every wild animal in the United States is, technically, owned by the public and managed by the government. That responsibility falls largely on wildlife agencies formed in the late nineteenth and early twentieth centuries, when insatiable demand for meat, fur, and feathers fueled hunting that drove some species—most famously the passenger pigeons whose flocks once turned skies dark—to extinction and many more, such as white-tailed deer, snowy egrets, and American crocodiles, to the brink.

Sport hunting associations pushed for the first regulations. At the time, even having a hunting season was an innovation. Theodore Roosevelt, who founded the legendary Boone and Crockett Club and helped popularize the ideal of fair chase in hunting, became President of the United States in 1901; he would set aside nearly 360,000 square miles for conservation, including the country's first wildlife refuges. New laws constrained trade in wild animal bodies and protected most migratory bird species. The 1930 publication of "American Game Policy," overseen by Aldo Leopold, helped define the nascent discipline of wildlife management, which for the first time became a profession run by trained biologists.

Schools educated young wildlife managers, many of whom went to work for the agencies that oversaw hunting, fishing, and trapping. Sales from licenses and taxes on guns and ammunition funded those agencies. Guiding them was a set of principles: Wildlife was a public trust. Policy should be based on science. Animals should only be killed for legitimate purposes. Their bodies should not be bought and sold. And, unlike in Europe, where hunting and fishing were largely the domain of aristocracy, here they would be accessible to anyone.

All this is sometimes called the North American Model of Wildlife Conservation, a name introduced in 1995 by the biologist Valerius Geist. It "ultimately led to a continent-wide resurgence of wildlife at a scale unparalleled in the world," wrote Geist, biologist John Organ of the US Fish and Wildlife Service, and conservationist Shane Mahoney in an essay entitled "Born of the Hands of Hunters." Though critics argue that such tributes downplay the role of nonhunters, the North American Model certainly accomplished a great deal of good—but it also contains contradictions and blind spots that have only grown with time.

Though wildlife management would incorporate ecology's lessons and evolve beyond "the art of making land produce sustained annual crops of wild game for recreational use," as Aldo Leopold defined his discipline, hunting and so-called game species remain central to the mission. Even as the agencies even play an ever-growing role in conservation, hunting and fishing attract the lion's share of what funding is available. One of Fred Koontz's frustrations was that, by his estimates, a mere five percent of the Washington Department of Fish and Wildlife's budget was allotted for the recovery of 268 imperiled nongame species.

Of course these aims are not necessarily opposed: protecting wetlands, for example, helps vast webs of life, not only waterfowl sought by hunters. There are times, though, when managing for game runs counter to what many people consider to be conservation and also to their ethical values. The most obvious example involves predators. Well into the twentieth century, it was common for wildlife managers to encourage the extermination of wolves, coyotes, bears, mountain lions, and other competitors for the prey hunters wanted. That degree of antipathy has lessened today, but it's still common for predator populations to be regulated in order to swell populations of target species.

Is that a legitimate purpose for killing? Reasonable people can disagree. What about killing for sport rather than subsistence? People disagree about that, too, and about the legitimacy of trapping animals for their fur. To some people—and, as a rule, to wildlife agencies—these activities are fine so long as they do not jeopardize a population's viability. To others, the issue of population viability is secondary to the question of how individuals should be treated; in the words of ethicist Bill Lynn, the agencies are "not thinking about the intrinsic value of these animals as persons: sapient, social, in relationship with others. Which raises the bar of ethical thinking." Certainly a great many people—not only animal advocates, but many hunters and other conservationists as well—consider killing contests to be illegitimate, yet they remain legal in most states.

Several years before the vote on killing contests, Koontz convened a meeting of scholars, activists, hunters, animal advocates, and conservationists to discuss how wildlife management in Washington might change. Among the ideas they discussed was something that reformers have pushed for in many states, with limited success: seats on the panels of appointees who set state wildlife agency priorities and rules. These state wildlife committees are overwhelmingly populated by people with backgrounds in hunting, fishing, and sometimes trapping: so-called consumptive users of wildlife. Yet if wildlife belongs to the public, is it fair for one portion of it to have such an outsized voice in their management?

A century ago, when those activities were far more common than they are now, that didn't ruffle many feathers. Nowadays, nearly eight times more Americans identify as wildlife watchers than as hunters; many people simply don't relate to wild animals as a "resource that can be renewably and sustainably managed," as a report by The Wildlife Society, the professional organization founded by Aldo Leopold, put it. The report labeled a belief in animal sentience as an outright threat. "If the animal rights philosophy was to become law," it stated, "wildlife would no longer be property, and would therefore fall out of the public trust."

Concern for the public trust is understandable. The middle of an extinction crisis is a poor moment to lose the support hunters provide for conservation. Yet it's not difficult to detect a deeper friction, a clash between value systems—and to exclude widely-held public values from serious consideration while managing a public trust seems contrary to both public trust doctrine and basic democratic principles.

And so, when the Washington Department of Fish and Wildlife invited Koontz to join its own commission, he was delighted. It was a chance to embody the reforms that he and others had hoped for; a chance to represent excluded perspectives on animals and push for wildlife management that focused more on conservation and less on consumption. It would also prove far more difficult than he expected.

◆ ◆ ◆

KOONTZ WAS ONE OF TWO NEW COMMISSIONERS. THE other was Lorna Smith, an environmentalist and wildlife advocate. In the language of social science, their presence gave mutualists three seats on the commission, with five occupied by traditionalists. It was a reasonable proportion, given Washington's demographics, but jarring for a hunting community that had long dominated the commission and felt besieged. Adding to the tension was the fact that a seat reserved for the eastern part of the state, a conservative-leaning rural region where hunting was more common than in urbanized western counties, was left unfilled.

Several days after their appointment, Kim Thorburn, a member of the commission from the eastern city of Spokane, published an op-ed decrying claims that certain controversial practices—trophy hunting, trapping, using dogs and bait to hunt bear—were unethical. She was one of the two commissioners to vote against a ban on killing contests, though she was hardly a stereotypical hunting cheerleader: as a retired public health physician and avid birder who devoted herself to sage grouse restoration and neither hunted nor fished, she seemed like the archetypal nonconsumptive user. But Thorburn cited the commission's statutory mandate to maximize recreational hunting and fishing opportunities. "My mandate does not say anything about ethical hunting," she wrote.

Moreover, she saw the criticisms as part of a culture war fueled by ideologues devoted to abolishing hunting. Smith, who had once been a hunter, and Koontz insisted that they had no such intentions, but the culture-war language was apt, and Koontz would soon find himself in the crossfire.

Later that year he and Barbara Baker, the commission's other mutualist, led a committee in drafting an interdepartmental guide to conservation. They defined it as "science-informed actions to preserve the health and resiliency of natural environments, safeguard the intrinsic values of non-human nature, and provide equitable benefits to current and future generations of people and species"—a noble, wonkily anodyne statement immediately followed by an assurance that conservation is the foundation of hunting and fishing. A small storm ensued. In a *Northwest Sportsman Magazine* article about the blowback, an unnamed agency biologist called it "subversive agenda-driven paradigm-shifting prose." Marie Neumiller, executive director of a hunting-oriented conservation group called the Inland Northwest Wildlife Council, bemoaned "us versus them" language that embodied an outdated preservationist ideal.

"We are a part of nature, so we need to stop separating humans from nature and conservation policies," she told the commission. It was a surreal moment: a concept originally used to expand conservation's vision—to stop chasing romanticized ideals of pristineness and treasure nature where we live and work—invoked to denounce a perceived threat to hunting. "I know that people are part of nature," Koontz told me. "They just shouldn't necessarily be in the center."

All that was prelude to a vote on the controversial topic of springtime black bear hunting. The Department's biologists had recommended continuing the hunt, which accounted for between 5 and 10 percent of bears killed by hunters each year; Washingtonians who objected flooded the agency with comments. They felt it was unfair to kill a bear who had recently emerged from hibernation, hungry and weak, and that the chances of killing mother bears and leaving cubs orphaned were too high. Hunt supporters argued that, according to the biologists, the hunt posed no danger to bear populations. It would also prevent damage to timber—in early spring bears will strip bark from trees and eat the sugar-rich sapwood beneath—and other potential conflicts.

Koontz also criticized methods used by department biologists to estimate bear population densities and calculate hunt quotas, though it wasn't really about that. "This is really about values," he said. "If the hunt is recreational, the majority of Washingtonians don't want it." And even if it was not purely recreational, was protecting timber company profits a good reason to kill bears? People might be sympathetic when a small local operator lost trees, but couldn't multibillion-dollar companies accept the appetites of a few hundred black bears as a cost of doing business in nature?

Many hunters didn't care for a spring hunt, either, but the lines were drawn. Koontz recalled one prominent hunter who told him privately of their opposition to the hunt, and yet testified to the Commission in favor of it. They feared a slippery slope that would someday end in a total ban on hunting. To Koontz this was unrealistic: there was simply far too much public support for putting-food-on-the-table hunting. Nevertheless, it was a concern voiced again and again during the bear hunt debates, as was the accusation that objecting to the hunt was antiscientific.

"I don't think fear has a role in science in a population that's stable," said one commissioner, a former wildlife biologist and refuge manager. *Northwest Sportsman Magazine* would describe "overwhelming public sentiment based in no small part on emotions," though biologists said the hunt "did not amount to a conservation concern for the bruin population." To treat a bear as killable for sport, then, or as a resource unit rather than an individual, was scientific, and to believe otherwise was not. There was a flaw in this reasoning: both these positions are values, not scientific fact. The department's biologists had applied science to the question of how best to conduct a hunt—but whether to have a hunt in the first place was not a scientific decision.

A discussion guided by science might also have considered the ethical implications of research on black bear intelligence. They're not nearly so well-studied as chimpanzees or dolphins, but what studies there are—as well as a vast amount of anecdotal evidence—suggest a comparable sophistication. They possess extraordinary problem-solving skills, an arguably language-like system of communication, and complex matriarchal societies. It has been suggested that bears exchange favors across years, with individuals sharing fruits and nuts from their own territories in exchange for assistance in lean seasons. Science can't say whether such intelligence merits heightened moral regard, but it certainly informs the conversation—the deliberations of state wildlife agencies do not typically consider it.

The commission deadlocked at 4–4; because of a clause in the department's bylaws, that canceled the spring hunt altogether. Outrage ensued. The decision made national news. "Washington State Lost Its Spring Bear Hunt to Political Overreach—And It's Just the Beginning," proclaimed an *Outdoor Life* article that lamented an "abuse of power" and "favoritism for urban areas" that was "emblematic of anti-hunting efforts taking place nationally—including in your state."

Disagreements were clearly about more than hunting bears in springtime. They struck a nerve in a subculture that felt itself persecuted and misunderstood, and raised thorny questions about the nature of collective decision-making in a liberal democracy. What is the correct balance between majority rule and minority rights? When is it appropriate for the values of people concentrated in one region to be imposed upon people in another? Of course the split was not quite so neat as that, and some people in eastern and rural Washington opposed the hunt just as some urbanites supported it. But shades of gray washed out quickly in the public debate.

It's easy to say that a majority, especially a large majority—polls suggested that 80 percent of Washingtonians opposed spring bear hunting— should prevail, and in principle that might be true. In practice it meant taking power from, and imposing power upon, the people who once had it. And while only 4 percent of Washington residents identified as hunters, and far fewer as bear hunters, their sense of grievance resonated beyond that community. The North American Model of Wildlife Conservation offers little guidance on how to soothe these troubled waters.

Fraught as the spring bear hunt fight was, it was soon eclipsed by the fight over the Blue Mountain elk herd, one of ten in the state, whose population had fallen to 3,600 animals. This was the lowest figure recorded since 1991, and far below the department's target population of 5,500. The department's own research found that only one in 10 elk calves survived their first summer. Unusually cold weather and inadequate nutrition may have contributed, along with predation by wolves and bears, but mountain lions were the largest factor, eating more than 40 percent of all calves counted. Department biologists suggested culling mountain lions to help grow the herd.

Koontz and Smith resisted. Perhaps the target elk population of 5,500 was too high, suggested Koontz, reflecting expectations set at a time when predators had been mostly eradicated from the state and more habitat existed. "This may well be a case where the socially accepted number is higher than the biological number," said Koontz. The suggestion did not go over well. On a commission Zoom meeting, hunter after hunter testified about feeling disrespected. A tribal wildlife manager said Koontz was displaying his white privilege. The host of a popular outdoor radio show on ESPN called it "the single dumbest thing I have ever heard a commissioner utter in 30 years."

There was a whiff of predator-hate and culture-war politics to the backlash. "In saving one cougar, you are condemning the lives of equally beautiful elk and deer," said libertarian radio personality Dori Monson, bemoaning the intrusion of "squishy liberals" into the Department of Fish and Wildlife. The antiscience charge appeared again, although it was again a clash of values. Should landscapes be managed to maximize elk populations or to foster the ecological relationships of which large carnivores are a natural part, with mountain lions having as much claim to elk as humans?

Koontz told me of research on all the animals who benefited from elk carcasses left behind by mountain lions. Hundreds of species relied, directly or indirectly, on that landscape-scale distribution of bodies. Yet from the hunters' perspective, they too relied on the elk; and though some of their prey would end up mounted on walls, it was hardly a trophy hunt. This was not just a few people looking for excuses to kill bears. People hunted elk for food, and far more people applied for elk permits than received them. Biologists had already suggested hunting fewer elk, and hunters had gone along with it—but to no avail. Why should they stand in line behind mountain lions? A landscape with fewer predators and more prey was still a marvelous place. They cared about nature every bit as much as big-city conservationists did and helped pay for the stewardship of these lands, yet here they were being lectured about social carrying capacity by some guy from Seattle.

It was not an easy conversation in the best of circumstances, which these were not. The criticisms took their toll on Koontz. A genial, gray-haired man given to wearing slacks and flannel, the sort of person you'd expect to bump into on a Sunday hike at the local land trust, he had already been unsettled by online chatter suggesting that commissioners who voted against the hunt should be hanged and wondering where he lived. "I don't think they really mean it," he said to me, "but maybe they do." One day he looked out the window and saw a man in a pickup truck parked in his driveway. It was too much. A few days before Christmas he resigned. "We have largely lost the ability to have civil public conversations," he wrote in his resignation letter. "I hope that your future appointments to the Commission can be better politicians than me, as I am a conservation biologist at heart."

Nowadays Koontz works behind the scenes, offering advice on how the department could focus on biodiversity conversation. Imagine what would be possible, he sometimes wonders, if the vast machinery of wildlife management in the United States was focused as much on biodiversity as on wildlife resources. Yet maybe it isn't worth the battle, he told me. Attempts at reform might end up making matters worse; they could push wildlife management into a full-blown culture war, with animals the ultimate casualty.

Or maybe all this turmoil was inevitable, a series of growing pains that would result, years from now, after difficult compromises and conversations, in a new but equitable balance between perspectives that now clashed without resolution. In spring 2023 the commission voted to halt the spring bear hunt, and Washington's legislature approved $23 million for the Department of Fish and Wildlife's nongame biodiversity efforts, effectively tripling that funding. Change is neither easy nor predictable, but Koontz is hopeful.

◆ ◆ ◆

SO FAR THE EMPHASIS HAS BEEN ON WHAT WILDLIFE management should not be: how people should not hold killing contests, kill bears for sport, prioritize elk over mountain lions, and so on. What about a positive vision of what wildlife management *could* be?

For that, one can travel a few hundred miles north to the Indigenous communities of British Columbia's Central Coast. Their ancestors had already been there for at least 14,000 years, with oral histories dating back to when retreating glaciers revealed the mountainous coastlines and broad river valleys of what is now a temperate rainforest, when European explorers arrived in the late eighteenth century. Across the millennia they developed systems for managing hunts and harvests, and also values that guided them. The values, too, should be understood as innovations, honed by experience so as to allow both people and the nature around them to flourish.

The systems proceeded from the values. Foremost among these was a sense of relatedness to nonhuman beings. Like the ancestors of Kathy Pollard and Margaret Robinson in the Northeast, coastal peoples told of a time when animals could transform into humans and vice versa: tales that are easy to see as mere myth but that contain a moral of connection. Relations with other creatures proceeded from an ethic of reciprocity, of generosity extended and returned in a feedback loop of mutual flourishing. "We are all one and our lives are interconnected," reads one of the core truths recounted in *Staying the Course, Staying Alive,* a collection of teachings compiled from interviews with elders of the

Heiltsuk, Kwakwaka'wakw, and Haida Nations. "All life has equal value. We acknowledge and respect that all plants and animals have life force."

Coastal First Nations believed that using those creatures came with a responsibility to maintain them, which sounds similar to the language of contemporary wildlife management—but with a key difference. They were maintaining kin; they were not entitled to them, but understood harvests to be a gift, and believed that greed or disrespect would be punished by those creatures whose generosity had not been properly appreciated. By all accounts these values and the lessons learned in their application stewarded a natural world bursting with an abundance that is difficult to conceive today.

When colonists arrived, the lands and waters that Indigenous peoples had so lovingly tended were treated as a storehouse to loot. The sea otters were liquidated; the millennia-old cedars, firs, and spruce trees cut; the whales slaughtered. Abalone, cod, halibut, sea urchin, eulachon, herring, salmon: each of these species was obliterated in a centuries-long spree of rapaciousness. In the early twentieth century, the Canadian government outlawed traditional fishing methods—the weirs and fish traps that First Nations people used to selectively harvest their catch, allowing the largest, healthiest fish to go free and enrich the population; the kelp-strung sticks on which herring would spawn, allowing people to gather their eggs without even catching the fish—to protect newly established commercial fisheries. These fisheries would, in tandem with dam-building, mining, and logging that destroyed the streams where the fish spawned, nearly destroy the very fish whose abundances were central to coastal cultures.

This was not simply a material hardship, but a fundamental reordering—a hollowing out, a degradation—of daily life. It was part of a larger policy that amounted to genocide, inseparable from displacing First Nations peoples from their homelands and forcibly sending their children to residential schools where their languages were forbidden, sexual abuse was rampant, and those who died from preventable diseases were often buried in unmarked graves. When First Nations people entered a period of cultural and political resurgence in the early twenty-first century, obtaining formal recognition of these wrongs and asserting their rights of self-governance, they also reclaimed wildlife management.

Grizzly bears figured prominently in this. While Coastal First Nations peoples had traditionally felt a kinship with all animals, they felt a particular closeness to grizzlies. The bears were seen as close relatives and teachers, "requiring respect and instilling an imperative of responsibility more in line with that felt by settler cultures primarily toward human relatives," wrote researchers describing a grizzly bear monitoring program launched by the Heiltsuk in 2006. From bears their ancestors had learned to nourish beloved berry patches with salmon carcasses, rub spruce pitch on their wounds, and eat the rhizomes of skunk cabbage to flush out intestinal parasites.

This kinship with bears, and in particular grizzly bears, was nearly universal among the people who shared their range, something many anthropologists remarked upon but that a study of plant names illustrates most poetically. Eurasian languages contain more than 1,200 plant names derived from bears: bearberries, bear's garlic, bear's cabbage, and so on. Some ostensibly gained their names for the plants' resemblance to bears, but many did because people had learned of them from bears. Coastal First Nations peoples spoke to grizzly bears with the expectation that they would be understood.

A recent analysis of the bears' population structure revealed three genetically distinct groups, their geographic distribution overlapping almost perfectly with the region's three Indigenous language families. It's not exactly clear why this is, but the most likely explanation is a mutual reliance on the same watersheds and food sources. Whatever the reason it's a fitting testament to the connection that people felt with them. And though colonialism's ravages eroded many traditions, as stories and practices died with those who remembered them, or withered through lack of practice, the kinship with grizzlies remained.

Hunting was common in these communities, but grizzly bears were off limits. That grizzlies were—with full government support—trophy-hunted within their homelands, their bodies left in skinless, headless, pawless piles, was a source of grief and a state-sanctioned desecration. In 2012 the Central Coast First Nations joined to oppose trophy hunting of grizzlies; they announced that, under Indigenous law if not yet under Canadian law, trophy hunting was banned, and in 2017 the British Columbia government agreed. They also worked with the Raincoast Conservation Foundation to purchase commercial hunting rights within the Central Coast—effectively ending trophy hunting of other species as well—and to conduct their own scientific research on grizzly bears.

These efforts had started years earlier, led at first by the Heiltsuk, who enlisted Raincoast researchers to help conduct a demographic analysis of the bears in their home Koeye Valley. They wanted this knowledge, wrote researchers led by Heiltsuk resource manager William Housty, not to exploit bears but to sustain them. They insisted that scientists treat the bears respectfully. There would be no radio collars to track their movements: how would *you* like to be knocked out and wake up to find a heavy metal collar stuck around your neck? Nor would researchers be allowed to remove their teeth, a customary method for assessing age. Instead they would rely on trail cameras, hair-gathering posts, and field observation. They would take seriously those people whose knowledge—and values—had been disregarded by government biologists. They would engage the bears as equals.

Ultimately this research led to a fine-grained understanding of where grizzlies lived in their traditional homeland. The Heiltsuk Nation controlled only a portion of this region; much of it was on so-called crown land, owned and managed by the Canadian government in consultation with the Nation, and used for commercial forestry. Within this land some of the bears' habitat was protected from logging, but not all of it, and the Heiltsuk now had the data to push for even more. Where logging and bear habitat overlap, den sites and the area around them are off limits—and that includes black bear dens, too. Rather than wanting them killed for eating tree bark, as wildlife managers in Washington had proposed, the Heiltsuk welcomed their presence.

Other Central Coast First Nations pursued their own bear projects. The Nuxalk Nation, whose home in the Bella Coola Valley was a hotspot of bear–human conflict during the spring and autumn salmon runs, formed the Nuxalk Bear Safety Group. Its members help install electric fencing, motion sensors, and spotlights around the fruit trees, compost piles, and smokehouses that lure bears into trouble; they monitor bear locations and share them daily on radio and social media, so people know where extra caution is needed. When people report that a bear has come too close, Bear Safety workers show up in trucks mounted with loudspeakers and flashing lights to help move the bear away. If a bear doesn't move, they stick around to warn people away from the site. It is the literal opposite of killing bears to reduce conflict.

The Kitasoo Xai'xais, further up the coast, focused on building their own ecotourism industry. The Spirit Bear Lodge—named after the rare subspecies of white-furred black bears who live in the region—is their town's second-largest employer. Meanwhile the Wuikinuxv Nation, who had witnessed firsthand the privations wrought on bears by declining salmon populations, decided to figure out how best to share salmon with grizzlies.

◆　◆　◆

BEFORE COASTAL SALMON RUNS COLLAPSED, BE-tween two and six million sockeye passed each summer through the short river that separates the silty estuary of Rivers Inlet and Owikeno Lake. ("Wuikinuxv" is pronounced much as the lake's name appears: Wee-Kee-No.) There they arrived after having spent their adulthood at sea, feeding on the zooplankton and small fish whose nutrient bounty they now carried to the streams of their birth, their muscular silvery bodies turning a deep scarlet as if to commemorate the many-thousand-mile swim that would end with their spawning and death.

Wuikinuxv families harvested them; they smoked their bodies on cedar stakes over long fires or in the eaves of smokehouses, processes that took days and provided the community with enough to last them through the coming year. The salmon were central to their life, just as they were for grizzly bears who gathered by the nine salmon-bearing rivers and streams that flowed into Owikeno Lake, eating the fish whose bodies would sustain them through winter's hibernation and fortify the milk that nursed their cubs.

Then came colonization: the unrestrained taking that would pinch this life-giving flood to a trickle. By the early twentieth century, more than a dozen large-scale commercial canneries operated in the inlet. They processed millions of fish harvested annually by hundreds of small gillnetting boats; their steam-driven machinery ran on local logs. The last local cannery closed in 1957, supplanted by industrial logging and more modern, mechanized commercial salmon fisheries. Enough sockeye remained for the Wuikinuxv—their own population a fraction of precolonization numbers—but just barely enough for the bears. Then, in the 1990s, the full toll of fishing and habit destruction came due. Salmon populations cratered. In 1999 a mere 10,000 sockeye returned.

It wasn't enough for the Wuikinuxv, much less for the bears. The bears starved. They wandered the streets of the village in search of food, fur hanging from their emaciated bodies, pressing themselves against windows and even breaking into homes. Mothers stopped defending their cubs. Nobody had ever seen anything like it. These were people accustomed to coexisting with bears, who might pass one on the way to school, or find one sleeping under a porch, and not consider it a big deal. What was happening was both frightening and heartrending. Some people wanted to feed the bears. They saw them as victims of a broken system, not so very different from them, in a way, impoverished by the greed and short-sightedness of others. Yet that risked creating a dangerous codependence, and the townspeople made the traumatic decision to shoot them.

The job fell to the village's most skilled hunters. Some still cry about it today. They asked for forgiveness; these were mercy killings, but still. They were killing their *relatives*. "Speak to the bears in our language," Jennifer Walkus remembers her grandmother telling her, "because the bears taught it to people." Walkus, who was born in 1971, well into the collapse but early enough to glimpse what life had been like, was the fisheries manager at the time. Even before that awful summer people knew that bears were not thriving as they once had, she said, that their cubs were born smaller than in the days when streams ran red with spawning sockeye, but now the full extent of their predicament was made viscerally real.

"The people who were living here at the time had seen so much of the suffering those bears had gone through," she said. They started thinking about how to ensure that the bears had enough fish to survive. The Wuikinuxv have a term for this: ṅàṅakila, to look ahead, to watch out for someone. That value would guide the scientific study of how to ensure that grizzlies had plenty to eat. And they would not be asking scientists from the government—the same government whose biologists supported the trophy hunting of their kin, who had allowed the fishing industry to decimate the salmon and the timber industry their habitat, who didn't take their own expertise seriously—for help. "They're the ones with the loudest voices," said Walkus of the industries. "The only ones speaking for the bear and the fish and the eagles are us."

Walkus went to Chris Darimont, science director at the Raincoast Foundation, who had worked on their collaboration with the Heiltsuk. He sent them Megan Adams, an idealistic young ecologist working toward her master's degree. She would help take Walkus's questions out onto the landscape. How many grizzly bears lived in their home? How much did they eat? How many salmon did they need, and how could their own salmon harvests include the needs of the bears?

The job was perfect for Adams. Rejecting the legacy of colonial dominion was important to her. She still remembered a painting a professor showed in one of her undergraduate classes, *American Progress*, John Gast's oft-reproduced 1872 celebration of Manifest Destiny, with "all the Indigenous people and bison and wolves fleeing the angel of destiny"—actually Columbia, the angelic personification of America, floating westward above the prairies—"trailing a railroad and telephone wires from her hands, and all these Pilgrims following peacefully behind her." That was the essence of colonialism, believed Adams, in Canada as well as America. It was still felt in the objectification of nature as a resource and the denial of personhood to animals and Indigenous people alike.

As an undergraduate she had studied salmon. Now she turned her attention to bears, setting up hair-snagging posts where grizzlies were likely to rub against them and analyzing the hair's composition to determine how many bears there were and how much salmon they ate. For some, the fish composed nearly two-thirds of their diets. Next they plugged the data and Wuikinuxv records of salmon runs into models that told them how different catch levels would affect bear populations.

This was an unusual step. One might think that sustainable fishing limits are already calculated with the needs of other creatures in mind, but such is not often the case. Sustainable limits typically refer to the number of fish people can take without causing their population to collapse; other creatures who also rely on them are rarely considered. Even so-called ecosystem-based fisheries models, a recent innovation intended to fix that shortcoming, often prioritize interactions with another species we'd like to consume—if X sardines, then Y tuna?—rather than sharing with the whole community of life.

The work took Adams and Wuikinuxv field technicians and other researchers brought in to consult the better part of a decade: thousands of hours of hiking the backcountry, counting salmon in streams, and crunching data, all to arrive at a simple figure. If current salmon populations held steady—around 200,000 sockeye now return each year—and the Wuikinuxv caught no more than 45,000 fish, or roughly 10 percent below what had previously been calculated as their sustainable limit, grizzly populations would be about 10 percent smaller than if they caught none. This seemed like a fair balance.

The bears' needs would also come first. Right now, said Walkus, talking to me in spring, not long after the salmonberries—named for their orange-pink coloration—had bloomed, the fishery remains closed until 100,000 sockeye have passed the counting station in the village. Only then is fishing opened to the Wuikinuxv Nation's several hundred members. For now, said Walkus, they take a small fraction of their newly calculated limit; they stay well on the safe side of what the runs can support, though hopefully a day will come when that number rises. But even then the bears will be first in line, followed by the Wuikinuxv, and it's unlikely that there will ever be enough fish to support a commercial fishery. "That's the scientific argument that we're going to use to keep the feds away from opening the fishery," Walkus says. "There's not even enough fish for the system to sustain itself. It's barely enough fish for people and bears to share. But there's enough. Leave it alone."

One might say that Walkus and the Wuikinuxv are being generous. Megan Adams had also used that word to describe salmon and herring. "I couldn't believe how generous they were, and how amazing they were," she told me. "These little fish, they leave the river this big"—she held her hands a few inches apart—"and come home this big"—she held her hands wide. "They fertilize everything and feed everybody and everything just explodes when the salmon come in. We have two very generous fish who come once a year. The herring are here right now, and then the salmon. Right after the winter and right before the winter, these fish come. And it's like magic. They're bringing protein and fat that wasn't here before. They go out and they sequester all this energy from the open ocean. And then they come home and leave it here in foot-high piles of herring eggs that the birds can eat for a month, or in an endless supply of fish to smoke and dry, and still generate millions of juveniles who can go out and do it all over again."

It's not only bears who benefit. Implicit in the model of how much salmon bears need are the seals, cetaceans, and eagles who come first, eating their fill out in the estuary before the fish have even entered the river that runs past Oweekeeno. And once they reach the streams where the bears wait to catch them, the life stored in their bodies ripples out through the landscape. The sheer scale is difficult to conceive: in an average year before the collapse, roughly 45 million pounds of sockeye surged into Owikeno Lake's tributary streams. Nowadays it's only a trickle of that, but even the trickle is a bounty. Shrews and mink feed on salmon carcasses left by bears, as do coyotes, bald eagles, and owls, some 80 vertebrate species altogether. Invertebrates feast, too: each carcass becomes a city for burying beetles and the larvae of flies who go on to help pollinate the flowering plants whose very growth is so nourished by decomposing bodies and bear scat that their leaves have unusually high densities of stomata, the structures that allow gas to flow between plants and the surrounding air. Plants breathe better when there are plenty of salmon and bears to distribute them.

The composition of plant species also changes. Because a wider range of plant species are adapted to nutrient-rich than nutrient-poor soils, bear- and salmon-enriched systems are especially bountiful and biodiverse. These systems are also rich in fruit-bearing plants that require animals to spread them. A single grizzly bear scat might contain hundreds of thousands of berry seeds: a winter larder for mice and voles, and a self-fertilizing plot for the next generation of plants. "We know that the more fish there are, the more bears there are. And the more bears there are, the more berries they eat, and the more seeds they shit out. There's reciprocity everywhere," said Adams.

◆ ◆ ◆

WALKUS RECENTLY STARTED ANOTHER PROJECT TO help grizzly bears. Many old Wuikinuxv village sites have orchards, no longer maintained, of crabapple trees that ripen around the time that sockeye return, and reviving them could nourish bears in years when salmon runs are thin. She brought on another young scientist, Sara Wickham, who is studying traditional practices for tending crabapple trees—pruning is yet one more skill the Wuikinuxv's ancestors may have learned from bears, whose snapping of shrubs as they fed yielded more growth in subsequent years—and bringing those old orchards back to life.

Decades from now those orchards will be a legacy of managing salmon and grizzly bears as kin. They will embody a vision of humans and nature as intertwined. But instead that state of being serving as justification for elbowing to the front of the line and taking even more, it's an obligation to step back, to share, to tend orchards that buffer vulnerable beings from the vicissitudes of fate. This isn't to say that the Wuiknuxv example offers a panacea or that every tension has been resolved: one wonders, for example, at what the pressure of a commercial fishery—that is, the voracious appetites and perverse incentives of global markets—would do.

Walkus thinks this is unlikely to be a problem. There are simply too few salmon and, if that were to change, the Wuikinuxv would resist the temptation to take more than is fair. She points to forestry as a case in point: their logging plans forsake short-term profit maximization for the sake of long-term ecological richness. Unlike the companies that can move on once a resource has been drained, they live here. And it's probably unfair, or at least unrealistic, to demand a system that is structurally incapable of having its purposes betrayed. The best we can hope for is that traditions and values will prevail.

But what about the practices of subsistence hunting and fishing? How is it possible to think of animals not just as resources—a term that coastal First Nations people do use—but also as fellow persons, as kin, and yet kill them? When it comes to returning sockeye, this arguably isn't such a dilemma—they spawn only once and then die, so by the time they pass Oweekeeno their lives have but a few days left regardless of what people do—but that isn't the case for other animals, including black bear, who unlike grizzlies are hunted. Some relatives are closer than others.

That's not an objection I personally would ever make to a Wuikinuxv hunter. After several centuries of genocide, dispossession, and being told how to live by outsiders, it would be the height of arrogance. The community's remote location—inaccessible by road, with the nearest store a floating outpost an hour away by boat—means that food imports are sporadic and expensive. But while the ethical questions posed by hunting may be avoided in that particular set of circumstances, they do remain. In my own life, I now try to avoid consuming any animals at all; I've certainly eaten far too many ever to be judgmental about it, and I still make exceptions, such as when someone offers me a meal they've worked for, but it's not something I'm comfortable with.

When is it okay to kill? To put an animal in the freezer is certainly preferable to killing one for sport, but that doesn't automatically make the ethical questions go away. I am skeptical of rationalizations often offered for hunting: the ethic of taking only those animals who have given their "permission," for example, seems self-serving and open to abuse. That the animals at least lived good lives, unlike those on a factory farm, does matter, but it's still tragic for those lives to cease. As for the notion that hunting is sustainable, there are simply not enough terrestrial vertebrates on Earth to support human appetites were farmed animals replaced on our plates by their wild kin. Their consumption might be sustainable in certain places, as on the Central Coast of British Columbia, where roughly 4,000 people are part of an ecosystem spanning 8,500 square miles, an area slightly smaller than New Jersey. But most people don't live in those circumstances.

Yet I also know that many hunters, among them people I consider my friends, have an ethos of care and respect. I'm aware that the impacts of my own diet, though smaller than they once were, still include the destruction of animals and their homes. Practices I consider unobjectionable might someday seem unconscionable. I certainly can't claim enlightenment or absolute ethical certainty. And in a society where living ethically is a constant struggle, and where so many pressures conspire to pull people apart, I'd rather find common ground than differences.

Here I turn to something that conservation scientist and ethicist Francisco Santiago-Ávila told me: he prefers to think of veganism as an ongoing attempt to do better, to think of what one might do to reduce harm, rather than a binary divide between those who consume animals and those who don't. What matters most isn't perfection. It's being thoughtful and trying to improve.

Writing in the journal *Conservation Biology*, a group of biologists and ethicists contemplating what it means to treat animals as fellow persons opined that it did not necessarily mean abstaining from harming them altogether or that obligations to them are uniform. "Between perfectly equal moral status for all and categorical moral segregation of the few," they wrote, "lies a wide expanse where a more inclusive and contextual moral space can be explored."

What is happening on the Central Coast represents one such exploration. And the principles guiding people there can be an inspiration anywhere: a relationship to the nonhuman world that is motivated by reciprocity and a sense of kinship—an acknowledgement of personhood, a willingness not to claim the lion's share for ourselves—rather than the strict categorization of animals as resources to whom we're entitled. How that manifests may differ from place to place and people to people. What else might be possible? What forms might wildlife management take when it proceeds from a spirit of generosity?

11

The Invaders

TEN YEARS AGO, WHILE CAMPING IN THE SONORAN Desert, Erick Lundgren noticed something unusual: clusters of holes, several feet across and at least that deep, dug in a dry riverbed. Water pooled at the bottom. The holes were wells, dug by burros—better known as donkeys, but still called by their Spanish name in the American Southwest. They fascinated Lundgren. In an otherwise parched landscape, the wells were unexpected oases. He wondered what other creatures might drink from them.

Lundgren, who at the time was a young biologist doing grunt work as a field technician on conservation projects, looked in the scientific literature for accounts of donkey-dug wells. He found none, which was not surprising. In their native East Africa, donkeys are critically endangered, with fewer than 600 remaining. There are hardly any to study. And in places like the US Southwest, where donkey populations are descended from animals imported centuries ago to toil for Spanish colonists, most ecologists and conservation biologists—as well as land agency officials and wildlife managers and ranchers—generally considered them a scourge: invasive, destructive animals despoiling delicate ecosystems ill-adapted to their alien presence, overpopulated, in desperate need of eradication. Their hole-digging didn't deserve attention.

They "do not play a functional role in our existing North American ecosystems," declares The Wildlife Society in a statement emblematic of the conventional wisdom. They're an invasive species and thus "among the most widespread and serious threats to the integrity of native wildlife populations." Lundgren, however, was not someone to accept convention on faith.

That had not always been the case. As a nature-loving kid and then a conservation biology-loving teen, he had learned that native species are good and nonnative species—those present because, at some point in history, humans moved them outside their previous range—are suspect. Whereas natives had evolved together into an intricately tangled community of life, nonnatives were interlopers; their ecological relationships tended to be fewer, poorer, and less life-sustaining. Some, freed from whatever checks and balances existed in their original homes, were destructive. These qualified as invasive. Killing them to protect ecological health was justifiable, even necessary.

One day, though, as he told his father about the dangers of Amur honeysuckle, an east Asian shrub now widespread in the eastern United States, his dad said something that made Lundgren pause. "I wonder what they will become?" he asked, meaning: What ecological relationships might, with time and evolutionary pressure, form around these newcomers? That resonated with Lundgren. It exemplified the curiosity and open-mindedness he believed a biologist should have. Soon he came to think that the dichotomy between native and nonnative species was far messier than he was taught.

Before moving to Arizona, Lundgren worked on an island off the coast of a California where cats were being exterminated to protect endangered native birds and to reduce competition with foxes.

(The foxes had, in fact, been introduced to the islands by precolonial peoples roughly 6,000 years ago—long enough ago to escape nonnative designation and also to become a distinct subspecies.) When the contents of their stomachs were examined, though, the cats turned out to subsist almost entirely on mice, while foxes were much more likely to eat the endangered birds. On another of these islands, conservationists had eradicated nonnative pigs—but the island's resident pig-deprived golden eagles started eating foxes instead, necessitating the capture of the eagles.

The tensions and ambiguities Lundgren encountered on the island are more common than one might think. In Micronesia, for example, purportedly nonnative monitor lizards were long targeted for eradication, but they proved to be a native species who reached the islands thousands of years before humans. In Mongolia, genetic studies of endangered Przewalski's horses treasured as the last truly wild horses found them to be descendants of domesticated animals. Does it make sense to justify the killing of ostensibly nonnative mountain goats in Grand Teton National Park when, a few dozen miles away, they are accepted as native? And might nonnative species sometimes provide benefits to biodiversity, enriching webs of life in ways that a focus on nativity had ignored?

Lundgren had worked on a project in Hawaii, where native birds have been decimated by mosquito-borne malaria. In many forests nonnative birds now fulfill the ecological roles of their native counterparts, spreading seeds from which future forests will grow—yet the birds were killed as a matter of policy. Closer to home, in the Southwest, nonnative tamarisk trees were blamed for crowding out native riparian forests and the birds who lived in them, but tamarisks were simply better-adapted than natives to survive drought and human-caused water scarcity. Endangered southwestern willow flycatchers now depended on them.

Moreover, beyond his misgivings with newly murky rationales for killing some animals to help others, Lundgren was troubled by how it was done. It wasn't that he couldn't stomach death; he'd grown up hunting. But this sort of killing was so casual and dispassionate, as if being labeled invasive exempted those creatures from even the barest regard. "If life doesn't matter, what the hell is conservation about?" Lundgren told me. Attempts at talking to his colleagues about this were met with stares.

By the time Lundgren decided to enroll in graduate school to study the interactions of native and nonnative species, these tensions had made their way into conservation discourse. "The practical value of the native-versus-alien species dichotomy in conservation is declining, and even becoming counterproductive," wrote scientists in a landmark 2011 article entitled, "Don't judge species on their origins." This remains far from a universal sentiment, but it was now part of the conversation. Simultaneously, the nascent compassionate conservation movement argued that the value of individual animal lives, and the ethical frameworks developed to make sense of how people treated captive and domestic animals, deserved a place in conversations about wild animals as well. Lundgren was a receptive audience for this, too, and it fed his discontent with reflexive nativism.

Though Lundgren's graduate work started with seemingly esoteric studies of nonnative grasses, the question animating him was profound: What might people learn if they set aside preconceived notions about nativity and nonnativity and if they looked with clear eyes at what these species actually did? He was, in short, the perfect person to notice the donkey wells.

◆   ◆   ◆

THIS IS HOW WE ENDED UP IN CATTAIL SPRING, ABOUT 140 miles northwest of Phoenix, the last few of which were traveled over a dirt track rendered nearly impassable by washouts during the rainy season. Cattail Spring is not a town but simply a place in the desert where water sometimes flows, and there Lundgren had placed trail cameras—the first of these he purchased with a crowdfunding campaign on IndieGoGo, underscoring how unconventional the research was—in the hopes of documenting donkey well life.

We arrived in late afternoon and made camp. As night fell, we talked about his frustrations with the dogmas of nativeness. He asked how I felt about eradications of introduced rodents on islands where they prey on seabird colonies. These are probably the most clear-cut cases of nonnative species having calamitous impacts on biodiversity, in which my own unease at killing some animals to protect others is overcome by an appreciation of those teeming colonies and the offshore reef ecosystems nourished by their feces. "That's the question I have a hard time with," he said. "Eradication on continents"—where populations are not neatly contained, and immigrating animals can quickly replace those who are killed—"I think is total bullshit. But islands are tough."

For my own part, I wish the killing of those rats and mice were at least accompanied by a sense of what environmental ethicist Chelsea Batavia and ecologist Arian Wallach, a prominent compassionate conservationist who was Lundgren's PhD advisor, called "the moral residue of conservation." It's not the rodents' fault that humans so heedlessly moved their ancestors around the globe; their appetite for seabird chicks would, if expressed by an acceptably native animal, be treated as an inevitable part of nature. To kill them, even for noble purposes, is to take innocent lives. "Conservationists should be emotionally responsive to the ethical terrain they traverse," argued Batavia and Wallach in the journal *Conservation Biology*. "Feelings of grief are commensurate with acts of harm. Apathy or indifference is not."

In all my years of reading and writing about the killing of invasive species, I've yet to encounter an expression of grief. To Batavia and Wallach, this is troubling because those feelings "act as tethers to abiding notions of what is good and of value in the world." To turn them off—Lundgren recalled a colleague who cried after euthanizing a native bird with a broken wing but killed nonnative birds with barely a change in expression—risks harming something important in ourselves. Callousness can only be maintained at the cost of compassion.

Lundgren agreed with this. A casual attitude toward killing introduced species, he added, also made it easy to avoid less tractable but equally important problems, such as the overfishing that is now starving many seabirds. Moreover, even on islands, the impacts of nonnative species could be nuanced: an analysis of 300 Mediterranean islands containing both seabirds and invasive rats found that rats limited the abundance of only one seabird species, something the researchers called "an amazing conservation paradox."

"We don't give any credit to evolution," Lundgren said. Perhaps, over time, newly introduced and long-native species would surprise us with their ability to coexist. Perhaps in many places they already *were* coexisting—but the ease of killing so-called invasives, and the habits of mind that reinforced, made it hard to see. I fell asleep to such thoughts beneath a starscape that, in the dry desert air and the absence of human habitation for miles in every direction, was as clear as any I'd ever seen.

At some indeterminate point in the night we were awakened by a nearby donkey's braying, a sound that is easily lampooned—hee-*haw!*—but in that place was nothing less than majestic. Proud, even, anthropomorphism be damned, and fiercely independent. At daybreak we found hoofprints: a group of donkeys had gathered a few hundred feet from our cots. Perhaps they had been curious about our presence; perhaps the bray was someone's way of letting us know that this was their home. From a ridgeline we sipped coffee and watched the sun rise, bathing the valley below in the gentle glow that precedes the crucible of day. There were no donkeys in sight but their distant brays echoed off canyon walls and mingled with coyotes howling their night's end.

"That's the sound of the Pleistocene," said Lundgren. By this he meant that until about 12,000 years ago—the end of the geological epoch more conversationally known as the Ice Age—the region was inhabited by a whole menagerie of large-bodied animals, among them camels, tapir, mammoths, and two species of wild horses, including one with an uncanny resemblance to donkeys. Like most of North America's large animals they were extirpated or driven extinct by a combination of climate change and human hunting.

Where megafauna still survive in their native ranges, though, they're considered landscape engineers and keystone species whose ecological functions are uniquely tied to their size. Research on them abounds: they disperse seeds across vast distances, distribute enormous quantities of nutrients in their dung and eventually their corpses, and create landscape-replenishing vegetative mosaics with their grazing. Through a long historical lens, then, their absence in the modern Southwest is both an aberration and a loss. To think of landscapes lacking them as being intact is a sort of ecological amnesia.

To Lundgren, the automatic perception of donkeys in terms of damage and alien origin reflects a failure to appreciate what megafauna do and a blindness to how conservation's idealized baselines are actually missing an entire guild of creatures. They're not identical to horses who once lived here, but their presence is far less strange than their absence. Lundgren identified a creosote bush for me. Creosote reproduces both through seeds and by cloning themselves; the oldest known colony is estimated to be nearly 12,000 years old. "It was one of the first plants to arrive at the end of the Ice Age," he said. "It germinated with mammoths and camels around it, but it's still here."

◆ ◆ ◆

AFTER LOADING OUR PACKS WE SET OFF FOR THE springs. Our footsteps crunched on the rocky, mostly bare soil, though it was hardly barren. The Sonoran Desert is one of the more biodiverse deserts on Earth, and as my northeastern eyes recognized few of the plants, Lundgren helpfully identified them for me: Crossosoma, prosaically known as ragged rockflower. Ocotillo, also called coachwhip, covered in spines that hinted at an evolutionary history of being eaten by large herbivores. A small yellow palo verde tree that had a very recent history of being eaten. "Look at what burros did to this plant!" Lundgren exclaimed. They had somehow reached into branches that stretched well above our own heads and pulled them down, snapping them off the trunk. "How do you interpret that?" he continued. "It's hard for that plant—but that's what herbivores do. It's unique, though. Nothing native is doing this now. But they *used* to do things like that."

A few of the breaks were fresh; others, weathered. We were not looking at a single episode of feeding but instead many spread across the years, shaping the tree's growth and, Lundgren suspects, effectively pruning the tree and stimulating the growth of tender new shoots that might be consumed by such animals as bighorn sheep. "The question is, how do you approach this?" said Lundgren. "What lens do you take? There are slides from Arizona Game and Fish"—the state's wildlife management agency—"that show images like this. The argument goes that native herbivores don't have front incisors, but wild donkeys do. Therefore plants die. Which I think is kind of magical thinking. Plus they're not dying. They're getting browsed."

Even so, I countered, wouldn't these trees be more vulnerable to disease? Every browsed tree might not die, but more would than in the absence of donkeys. "Totally. That's possible," Lundgren answered. "But what does that mean? Is that good or bad? I have a hard time if a system changes from having topsoil to having no topsoil. But shifting the species composition? Or shifting the architecture of trees? Those are not clear-cut good and bad, I don't think."

He pointed out an untouched palo verde tree not far from the first. "You can even make arguments about the patterns," he said.

With one tree stunted by browsing, perhaps the other would be freed to spread wide and become a canopy tree. He directed my attention to how the donkeys consumed only part of the branches they harvested, leaving the twigs in a thorny mat at the tree's base. This is common, he said, and he mused about what functions the pile might perform. Perhaps it would allow grasses to become established without being grazed by cattle; at a glance, those stems poking from the pile indeed seemed taller than their neighbors. Beneath the twigs the soil was comparatively cool and moist, all the better for seeds to sprout. And what other organisms took refuge there? Might it be considered a unique microhabitat?

"I would love to know what happens to this blanket of thorns," he said. "You don't have that in the environment otherwise." Maybe, he admitted, he was wrong—but it was the sort of question that tended to be dismissed without investigation by virtue of the donkeys' nonnativity. Yet if someone didn't know the donkeys' origins, would they still consider their activity to be damage? If they thought that donkeys were native, would they look at this in an entirely different light? Lundgren liked to think about submitting a study, tongue-in-cheek, in which he enumerated all the activities of a large, disturbance-creating native creature, such as a moose, without identifying the species, implying that he was discussing the destruction wrought by an intruder—until revealing that all along he was actually talking about a native.

We continued our walk. At one point we crossed a track that Lundgren identified as belonging to javelina. Pig-like peccaries who are historically native to South America, they gradually migrated north, arriving in the United States during the twentieth century. They're now considered a native species, yet donkeys have actually been here longer. If donkeys are problematic because modern ecosystems didn't coevolve with them, said Lundgren, shouldn't javelina also be viewed with suspicion? "If an animal managed to get here on its own," he said, "then even if it's going to have the same effect as an animal we brought, it has a right to be here. But the animals that we brought as slaves and dumped, don't belong."

Several times we passed donkey poop: not the midden piles they're known to make, rich with information about identity and territory, but small desiccated cakes. "There has to be some ecological implication," said Lundgren. "Follow this place for the next 20 years and see what happens to it. Or just in terms of the termites and dung beetles bringing carbon into the soil." Indeed that is one of the subtle legacies of megafauna loss: the world's dung beetles, those humble, little-celebrated, subtly important cyclers of fecal nutrients and nourishers of soil, are one-third smaller than they were when megafauna collapsed.

The scat was fibrous and hinted at yet another role for donkeys: plant propagation. Precious little scientific research exists on the ecological interactions of donkeys, but an exception comes from Belgium, where ecologists at the University of Ghent studied what happened when donkeys were allowed to graze in Flemish nature reserves in the hope of mimicking processes lost when native large herbivores went extinct. The seeds of no fewer than 82 plant species hitched rides by snagging on their coats or passing through their digestive tracts, aiding their spread throughout the reserves.

The long-distance nature of this seed dispersal is a crucial point. It allows genes to circulate through landscapes, preventing local plant populations from becoming isolated and stagnant, and helping species colonize new locales. The Belgian ecologists described the donkeys as "mobile link organisms." "Since large herbivores have always been part of natural ecosystems," they wrote, their role in dispersing seeds "is probably indispensable for maintaining species richness." Their findings were echoed by observations from central Iran, where researchers studying seeds in the feces of Persian wild asses—a close relative of African wild donkeys and their domesticated or feral descendants—said they "play an important role in plant biodiversity protection and vegetation dynamics."

Other researchers have concluded that, in the absence of megafauna—not donkeys specifically, but the guild to which they belong—the maximum dispersal distance of fruit seeds has contracted by two-thirds since the Pleistocene. Lundgren noted a Joshua tree, a beloved western species threatened by climate change. By some estimates up to 90 percent of them will be lost by the century's end; their habitable range will contract dramatically. It's hoped that the trees will survive by migrating north, but whereas their seeds were once carried for miles by camels, "the only thing that disperses them now are rodents," he said. Deer eat them, but chew the seeds so finely that they're extremely unlikely to pass through them intact. Might donkeys do what the camels once did? Could they help Joshua trees find refuge? That was yet one more question for a future study.

By this point we had walked for a few hours. I started to wonder whether we would actually see the subject of our theorizing. And then it happened: two loud snorts alerted us to the fact that we were no longer alone. Atop a low slope to our left, perhaps a hundred or so feet away, was a donkey.

◆ ◆ ◆

IMMEDIATELY WE STOPPED AND FELL SILENT. IT'S FAIR to say that the donkey startled us, rather than vice versa. He stood in profile, keeping one dark eye and a swiveling ear trained on us, and had almost certainly been watching us for a while. Perhaps the snorts had been his way of telling us that he knew we were there. I snapped a few photographs and he gave two long sniffs, breathing in rather than out this time, as if inspecting us through his enormous nostrils. Later on when I listened to my recording of the encounter I heard the sound, which at the time my ears missed, of a distant donkey's neigh, so perhaps he had been communicating. After a moment he returned to browsing a patch of brown grass, unconcerned but vigilant, raising his head every so often to check on us.

I reflected later on how I might describe the donkey. In strictly physical terms he had a sturdy gray body and muscular white legs; his muzzle was also white, as were the rings around his eyes and a crescent under his neck and his ears, which were fringed with a black that accentuated their alertness. But it occurred to me how little I could know about him. Not because a donkey's interiority is forever inaccessible to us or some such excuse, but because I was simply a passerby. Who were his friends and family? What did he enjoy doing? What was his story? It's so easy to think of wild animals in surface terms, to put the label before the individual, because we only glimpse them.

In the autumn after I visited Lundgren, though, I started volunteering at the Peace Ridge Sanctuary, a haven for rescued farm animals in rural Brooks, Maine. They started me with the donkeys, 43 of whom live there, a few rehomed by caring people who could no longer provide for them, but most were victims of abuse and neglect. On Sunday mornings I shoveled their stables; there I came to know a few, at least a little. There was a self-possession to those donkeys, a sense of thoughtfulness and reserve, that made clear how those brief visits only amounted to an introduction. If that assessment of their character sounds like a romanticized projection, I can only offer that it's shared by a great many people who know donkeys. At any rate the generalizations matter less than their individual stories.

There was Angie, who arrived as a foal with a dislocated leg that required her to be kept separate from the main herd until it healed. Between that and not having a mother to instruct her in social graces, she was an outcast at first and reminded me of children I'd known who simultaneously craved social connection but sabotaged it. She certainly had no reserve about seeking attention, almost invariably trotting over to meet me when I arrived. Her behavior stood in sharp contrast to Jed, who often stood near me but didn't come too close, looking at me with deep eyes that conveyed something like shyness, even a lack of self-confidence, leaving it up to me to stand with him.

The simple act of standing with them seemed important to the donkeys. They were not much given to unreserved affection in the manner of dogs or goats—although they did love a good cheek-scratch—but were quite social and liked to be near one another. Ethologists reviewing research on donkey cognition and behavior call this "dyadic proximity maintenance," a phrase both accurate and inadequate. Some had one extra-special friend whose side they rarely left, a phenomenon that is well-known to people who know donkeys and has also been confirmed by scientists. Though most explanations focus "on sexual and kinship reasons as the motivation underlying the pair formation," wrote the ethologists, "not rarely the members of the dyad are of the same gender and unrelated." That is to say, these bonds are not motivated by sex or kinship but by affection, which is why separation "may result in outstanding distress, inappetence and pining."

One such friendship involved a miniature donkey called Cinderella—thusly named, I suspect, because her eyes were so large and lovely as to give her a fairy-tale princess sort of quality—and Zinnia, who came to Peace Ridge by way of a notorious animal auction, the sort of place where horses and donkeys are dumped when they've outlived their profitability or convenience and are purchased by so-called kill buyers who sell them to slaughterhouses. A rescuer there took pity on Zinnia, who was pregnant and being picked on by the other donkeys. She didn't actually know how to care for a donkey, though, and so Zinnia ended up at Peace Ridge, where Cinderella immediately took her under her wing.

It was a good match. Until Zinnia arrived, Cinderella had been something of an outcast. And whereas Zinnia tended to be shouldered aside by the other donkeys at feeding time, Cinderella was a no-nonsense competitor who kicked off the others and allowed Zinnia to eat in peace. Zinnia was the only donkey with whom Cinderella would share.

Apart from jockeying for hay, there was rarely any conflict among the donkeys, and not much of a hierarchy. It was odd to think of Cinderella, tiny and sweet-eyed, dominating anyone else; were that purely a matter of size, then a gentle giant named Lucas would have reigned supreme. He was an American mammoth, a breed developed for size and strength, as tall at the shoulder as I am. His mother was slight-framed Mama Twilite, who at 27 years old was the oldest of Peace Ridge's donkeys. They arrived at the sanctuary together after state animal control officers found them starving and infested with lice in a lightless barn.

Mama Twilite had a hernia from a lifetime of pregnancies. She also suffered from Lyme disease, which caused her joints to be sore, and a pituitary disorder that lengthened her hair. Shaggy and slow-moving, she seemed to embody a certain patient, earthy wisdom, virtues that shouldn't be projected on the basis of appearance but seemed appropriate nonetheless. She was a favorite at the sanctuary and accepted the attention of visitors but rarely sought it out. One exception, though, occurred one afternoon not long after my cat died. He was like a little brother to me; I was still aching. I happened to walk past Mama Twilite, who was sitting in the grass, and sat down beside her. She put her head in my lap. I don't know if she sensed my pain and intended to console me, but it was the only time before or since that she was so affectionate, and it certainly had that effect.

After several years I still wouldn't presume to say that I know the donkeys at Peace Ridge. My brief visits—eventually I moved on to goats, and nowadays I just say hello to the donkeys on my way out—made me aware of how much there is to know about each of them. Each one had a story.

◆　◆　◆

IF I COULDN'T PRESUME TO KNOW AT A GLIMPSE THE donkey who Lundgren and I encountered, I could know that he had a story, too, and though I didn't know it, I could at least speculate about some of the historical forces that shaped it, beginning with the domestication of his ancestors by northern African pastoralists some 6,500 years ago. They were used for milk, meat, and most of all labor, particularly the transport of heavy goods across mountains and deserts. Their breeding soon spread to Mesopotamia, and donkeys became central to the economies of the ancient world; Romans would later spread them throughout their empire. Yet for all their importance, wrote Jill Bough, author of *Donkey*, a sociocultural history of the species, donkeys "have invariably enjoyed a low status in human cultures, received little appreciation and been treated harshly."

Unlike the horses favored by royalty, donkeys were a lower-class animal, symbols not of pride and nobility but instead objects of ridicule and derision. No fewer than 20 of Aesop's fables concerned donkeys, who in the words of Bough were "invariably depicted in a negative light, as servile and lazy, selfish and stupid; used as examples of folly and weakness." It's all the more poignant, then, that Christian lore depicted the pregnant Virgin Mary carried by a donkey into Bethlehem and Jesus atop a donkey when he entered Jerusalem. The Biblical tale of Balaam, who savagely beat his donkey for swerving to avoid an angel whom Balaam could not see and was reprimanded by God speaking through the donkey's mouth, suggests a regard for their humble, long-suffering wisdom. But that scriptural sympathy made little difference in their lives. *The Golden Ass*, a second-century fiction about a young Greek named Lucius who was transformed into a donkey, offers a glimpse of what these were like: he was used to carry heavy loads and to turn a millstone, sold multiple times, and tortured for his owners' pleasure. Medieval attitudes changed little, and it was

Shakespeare himself who took their name, ass, and popularized it as derogatory term that proved timeless.

In the sixteenth century Spanish colonists transported donkeys to present-day Mexico. Their desert-adapted hardiness and steady temperaments were useful for many forms of labor, most of all mining: they were forced to carry bags of rocks up mountainsides and through tunnels beneath them, work that was dangerous for humans as well—but at least humans could choose to leave.

Yet not all relationships were so brutal; decency and even a sort of solidarity could prevail. Prospectors and miners—whose image survives today in the trope of a grizzled man, pickax on his shoulder and saddlebag-laden donkey at his side, crossing a forbidding desert—often described donkeys as their closest friends. To them donkeys were more like Dabble, the steadfast, beloved companion of *Don Quixote*'s Sancho Panza. When in the late nineteenth century the depletion of mines and rise of mechanical transportation rendered their friends superfluous, these men allowed their donkeys to go free rather than killing or selling them. Other donkeys escaped. Together they survived.

That evolution had adapted them to arid environments would not have excused them from the desert's lessons. In that harsh, beautiful world of stone, thorn, and poison they learned the washouts and rock slides, the rattlesnakes and scorpions, and the locations of water on a parched landscape that guards its moisture fiercely through the long months when rain does not fall. For the first time in six thousand years of what can only be described as slavery, donkeys were born free. "Quiet and stupid as a burro usually appears," wrote one observer at the time, "he is quite another creature when free."

In the coming decades they learned how to avoid mountain lions, their last surviving natural predator in these defaunated landscapes, and also people with guns, of whom there were many. "Some people slaughtered burros for commercial purposes"—turning them into leather, pet food, and fertilizer—"while others did so for fun," wrote historian Abe Gibson in an essay on donkeys in the Southwest. "They all enjoyed the full support of ranchers, biologists, and management officials. These parties all agreed that feral burros were invasive pests who needed to be eradicated."

It appeared they soon would be, but aging prospectors made one final gesture for their old friends, campaigning on their behalf and securing at least some protection and public sympathy for the animals. They slowed the extermination, although well into the latter half of the twentieth century wild donkeys could be hunted without limit. State wildlife agencies encouraged their persecution. Donkeys "take over and ruin desert springs and water holes, destroy forage for game and stock, and are one of the few animals which will kill the young of other species," read a 1953 missive from the California Department of Fish and Game that referred to donkeys as "dust-caked jackasses."

Sometimes people ate donkeys, but often they simply found pleasure in killing. There was even a Burro 100 club with membership granted only to people who surpassed that number of kills. "Remnants of initiation to the club might be visible still in many areas in the form of burro 'grave-yards,' where groups of burros were shot in mass," wrote biologists in 1981. Lundgren described meeting one eminent mammalogist and telling him of his interest in donkeys. "He was like, 'You must just want to shoot them every time you see them,'" Lundgren said.

Their populations, and also those of wild horses, plunged to levels that alarmed their supporters. Thanks to the efforts of Velma Bronn Johnston, known to history as Wild Horse Annie, they were protected, finally, by the federal Wild Horse & Burro Protection Act of 1971—a rather incomplete, contested, and complicated form of protection that still allows for large-scale roundups and persecution, but at least through official channels rather than a free-for-all. People might sometimes shoot donkeys illegally, and in some places their extreme skittishness suggested as much to

Lundgren. But the donkey we met was reassuringly untroubled by our presence.

He could be said to have grown up in a brief interlude of peace. I wondered if his people had stories about us. When I voiced the thought Lundgren did not, as might be expected of a scientist, mock the notion. To his mind there was no doubt that donkeys had a society, and he wondered at the meaning certain places held for them. He mentioned a red rock butte not far away, a sort of giant pillar, at the base of which donkeys liked to gather and romp.* "Why there?" he said. "It doesn't seem like there's a resource that they're using. It's just a place for them."

Some of the region's Indigenous inhabitants, said Lundgren, once learned songs that were their birthright and inheritance; these were rhythmic and melodic descriptions of the landscape, both song and map. Might we find place in a donkey's long bray, too, if only we could translate it? And what of their customs, their culture, the society? Lundgren has observed that wells are dug by one individual while others watch, and youngsters will often stand next to the digger and go through the motions themselves, suggesting that well-digging is a learned behavior. Perhaps the techniques are refined and passed on, a part of local cultures unknown to us.

Every afternoon the donkeys at Peace Ridge walked their fields in single file, following a network of trails that after years of daily use were as neat as bicycle paths. What customs and rituals did this donkey's community have?

Even the basic social organization of donkeys is something of

---

* A study from Namibia describes how shallow depressions created by Hartmann's mountain zebras as they roll in dust gather more rainfall than the level ground around them. Eventually these depressions become tiny hotspots of plant and invertebrate life. Dust bathing is common in equids, including donkeys; it helps maintain their coats and is also a social ritual performed in groups. I sent the study to Lundgren, and a few months later he emailed to say that he'd observed similar vegetation hotspots where donkeys rolled.

an uncertainty to scientists, but there is a tantalizing complexity to them. In a year-long study of donkeys in the Bill Williams Mountains, not so far from where Lundgren first noticed the wells, zoologists Rick Seegmiller and Robert Ohmart of Arizona State University observed them in a variety of configurations: single males with multiple females and their young; mixed-sex groups of adults; single-sex groups of males or females and their young; and sometimes solitary adults. Individuals frequently left one group and joined another; at night groups mingled and split further as donkeys congregated at watering sites and fields. This might also be called a fusion–fission society, similar our own.

What do we owe them? This, at least: open-mindedness and a willingness to challenge our own assumptions.

◆ ◆ ◆

ALREADY THE APRIL MORNING WAS SUMMER-WARM BY my standards, my shirt clinging to my back, but this was comparatively balmy. In a few months the daily temperature would soar above 100 degrees Fahrenheit; in the empty trough between the so-called rainy seasons, when a wet month might bring an inch of precipitation, months could pass between rainfall. The red gravel that crunched underfoot made each step a reminder of how bone-dry it was. Even when water fell, there was little soil to hold it, and it coursed off the landscape through flash-flood channels or percolated into subterranean aquifers. Plants went dormant or guarded their moisture behind cages of thorns. That desert creatures have adapted to these conditions makes them no less harsh, and the distance between life and death may be measured in a few tenths of an inch of rainfall.

The spring was a sandy wash at the bottom of a small box canyon. Vegetation lined the canyon's sides, and a rustling above us pulled our eyes to a pair of cows watching us warily from a tangle of bushes into which they fled.

The presence of cattle was also evident in the cakes of dung that dotted the wash. This is one of their unfortunate habits, said Lundgren: cows will congregate around water sources for long periods of time and poop where they stand. Donkeys would leave a few deposits behind, but not in the overloading volumes that cows did, causing waters to become fouled by microbes feeding on the nutrients. By digging fresh springs, then, donkeys might help those creatures whose waters would otherwise be polluted by the cows— and, for that matter, they'd also benefit the cows, against whom Lundgren bore no enmity. They hadn't asked to be here.

Lundgren found five donkey wells altogether, some no more than moist depressions in the sand and others containing fresh water. Not very much fresh water, and in my mind's eye I had expected more, but even a small amount was precious in that parched landscape. Lundgren recommended that I think in terms of water webs as well as food webs, and in a landscape where many creatures acquired water not only by drinking but by consuming plants and other animals, a donkey's well might ease the pressure. Lundgren mentioned experiments showing that the amount of vegetation consumed by crickets and the number of crickets consumed by wolf spiders decreased in direct proportion to their proximity to water. Not only did water support life; it made competition less fierce, the struggle less brutal.

One of the wells hummed from a distance with bees and flies who had come to drink at its edge. As I approached, their drone was a wall of sound. Sometimes, said Lundgren, the pools were surrounded by clouds of butterflies, and he or his assistants could stand in the center of a butterfly vortex. Squirrels and warblers would drink beside the researchers as they worked.

Lundgren showed me a patch of cocklebur, a wetland plant named for the spurred seedpods that caught on the fur of passing animals. Perhaps they had arrived here on a donkey's coat or germinated from the fertilizer of their dung. Lundgren has observed donkeys browsing on mesquite seedlings that appeared to have grown from their dung; one mesquite species in particular, the screwbean mesquite, is considered especially important, with the seeds within its eponymous pods a vital food for dozens of animal species. Screwbean mesquites are also in decline across the Southwest, dying off for reasons still unknown at such rates that their range has contracted by half in less than a century and causing some scientists to worry of extinction. The donkeys effectively created nurseries for them.

I waited in the shade of the canyon wall while Lundgren checked his trail cameras, leaping from rock to rock with a surefootedness that, hackneyed as it was, made me think of donkeys. We didn't see the cows again, although on the way out we found the bones of one fortunate enough to die here rather than in a slaughterhouse. That's an uncommon fate, though; relatively few cows die on the range, and their nutrients are delivered onto our tables rather than circulating back through the ecologies where they're gathered. "I can't wait until condors show up on these landscapes with donkey bodies to eat," Lundgren said.

Months later Lundgren would recover the cameras and learn who had used those wells. The cameras didn't do well at detecting smaller animals, although it was at least possible to see Sonoran Desert toads in abundance. But they had no trouble detecting ringtail cats, striped skunks, javelina, a mountain lion, mule deer, bobcats, gray foxes, scrub jays, Scott's orioles, and even a passing kingfisher. There were donkeys with striped legs, a trait lost during domestication and now returning, as desert life had once again made camouflage useful.

Ultimately those photographs would become part of a dataset that contained 2.5 million photos collected at four sites across three summers. They were the centerpiece of Lundgren's PhD thesis, which would be distilled into several scientific journal papers. In one he compiled a list of traits—body size and shape, dietary habits, digestive physiology—for every herbivore species weighing more than 10 kilograms to have lived during the past 130,000 years. With those he quantified the properties of past and present herbivore communities, and in another paper argued that mixed-up modern herbivore assemblages, containing both native and nonnative species, are actually truer to the pre–mass-extinction communities of the Late Pleistocene than are native-only assemblages. And then, in the pages of *Science* itself—one of the world's most prestigious scientific journals, a place he could only dream of when crowdfunding for trail cameras—he described the well-digging donkeys.

There his speculations and early observations took their mature form. At each of his four Sonoran Desert sites, donkey wells provided water when streams dried up and summer temperatures peaked. They increased the total surface area of water on the landscape by between 74 percent and 100 percent, and the total amount of water available by a factor of 10. Having wells meant that the average distance between open water decreased by roughly a mile; no longer would creatures have to go so far to find moisture, nor compete so fiercely for it. Finally, he counted the vertebrate species using those wells and compared them to animals photographed at springs without donkeys. No fewer than 59 vertebrate species used the wells. By measures of species richness, visitors to the wells were a slightly more biodiverse group than those at the springs.

In the *Science* paper he also described how at one site the wells hosted a higher density of willow and cottonwood seedlings than did riverbanks. Willow and cottonwood are the backbone of riparian forests in the Southwest, and an urgent conservation concern; they require moist soil in which to germinate but struggle to take root amid existing riverbank vegetation. Lundgren suggested that conditions at the edge of donkey wells—sandy, moist, free of competing plants—actually mimicked those created by floods, a phenomenon diminished in a landscape of flow-regulating dams and agricultural diversion. Not only did donkeys spread seeds. They prepared the ground itself.

All this was translated to terms befitting scientific publica-

tion. "Our results suggest that equids, even those that are introduced or feral, are able to buffer water availability, which may
increase resilience to ongoing human-caused aridification," concluded Lundgren and colleagues. The measured phrasing belied
how radical a notion this was. Pushback soon followed: Researchers at the Arizona Game and Fish Department penned a letter to
Science calling Lundgren and colleagues' positivity "troubling."
Feral equids, they said, deprive native wildlife of food; exclude
other animals from water sources; reduce plant diversity; and, on
the whole, produce degraded habitats with a diminished capacity
to sustain life. The discussion was somewhat muddied by the fact
that Lundgren had also observed well-digging by a band of wild
horses. His *Science* paper subsequently referred in general terms
to the well-digging benefits of equids, horses and donkeys alike;
the research cited in the rebuttal focused exclusively on wild
horses. Those animals are even more controversial than donkeys,
and although the debates about both species have much in common, it wasn't quite fair to make a point about one by invoking
studies on the other. Yet in a way that underscored Lundgren's
point: people were quick to dismiss donkeys.

He and the advisers who guided the research fired back with
their own rebuttal. Yes, they conceded, donkeys and horses could
have those effects—but so could any native herbivore. "Describing
feral equids as either ecological heroes or as invasive pests over-
simplifies complexity and moralizes ecology," they wrote. Far
more important than their nativity was the context shaping their
impacts, particularly predation—and most research on wild equids,
including every study their critics cited, examined their impacts in
the absence of predators. It was like labeling all white-tailed deer
with the findings of studies conducted in suburban parks.

In fact, Lundgren had studied predation, too. In that study, he
found that mountain lions regularly prey upon donkeys. Their presence on the landscape caused dramatic changes in donkey activity;
they spent far less time near wetlands, making it easier for other

animals to access the waters and reducing their impacts. "Cougar predation appears to rewire an ancient food web," he concluded. But as cougars are hunted across their range, both for sport and at the behest of the livestock industry, that web only sometimes unfurls.

◆ ◆ ◆

THE DEBATE OVER DONKEYS' IMPACTS WON'T BE SET-tled without more data. But Lundgren's work has made their ecological beneficence at least a possibility, and it is already changing minds. I interviewed several researchers sympathetic to Lundgren, including an old-school Arizona State University ecologist named Dave Pearson. By his own admission he had long thought donkeys belonged nowhere except in captivity or the landscapes of their Ice Age range. "Burros anywhere else were to be shot out and gotten rid of," he said. Lundgren's work had helped change his mind.

"They have almost become a keystone species," he said. "Their presence makes it possible for other species to survive where they couldn't before." At high population levels they certainly could over-graze and strip their habitats, but "those situations are not all black or white." He thought the populations Lundgren studied did more good than harm. I pressed him on this point: perhaps donkeys could do good, so long as they had intact landscapes on which to migrate and predators to regulate their populations, but that was seldom the case. "That's possible too," Pearson said. "I would guess that, intro-ducing burros everywhere, you'd find some places they would do ter-rible damage—and other places where, as Erick's study shows, they would be very beneficial. Does that mean we should eradicate all burros? Or not touch any? We need to look at a case-by-case basis."

Lundgren himself chafes at the notion that the value of donkey lives depends on their proving themselves worthy. If it were found that elephants actually have negative impacts on biodiversity—and indeed, depending on the metrics by which they're judged, the impacts of their browsing and trampling can decrease the diversity of certain creatures—they would not suddenly be undeserving of consideration, nor any less valuable as individuals, he argues. This sensibility has come to nourish his own regard for wildness and wilderness. Those concepts are not, as has become the fashion in some academic circles, outdated values predicated on a misanthropic sense of human presence as antithetical to pristine nature. It isn't about that at all, he contests.

In an email, Lundgren described receiving a message from an ecologist who felt that he was woefully misguided. Lundgren fired back angrily; to his surprise they shared a fundamental agreement. "The common ground was the suggestion that we are managing the wildness out of the world, which seems to be the one thread nativist ecologists agree with me on," Lundgren told me. The very word wild, he had learned, came from Old English word that meant self-willed. "Suddenly, wildness, and respecting it, is about *rights*. The rights of the more-than-human, and human, to be self-willed. Wilderness then becomes a cultural landscape where the rights of humans and more-than-humans are respected."

He wrote that email after a painful trip to Death Valley where he found that the donkeys he studied for years had all been removed as part of the US government's horse and donkey management. The policy is intricate and the arguments around it correspondingly so, but the upshot is that the government wants none to live in national parks, and on other public lands it has pledged to reduce donkey populations by roughly two-thirds from their current 16,000—a process accomplished by rounding them up with helicopters and keeping them in pens until they are sold.* Thanks to campaigns by

---

* Painful and deadly as this may be, it's at least far preferable to a method frequently used in Australia, where so-called "Judas donkeys" are captured and outfitted with location-transmitting radio collars. Once freed they seek out their friends and families, who are then gunned down in front of them, again and again, until only the Judas donkey is left.

animal advocates they're at least supposed to end up in decent homes rather than slaughterhouses, although that doesn't always happen and some die during capture or are sold into slaughter. One can also imagine the trauma of having families and friends torn apart.

The possibilities raised by Lundgren's research are not considered in this policy. As became clear as Lundgren and I drove back to Phoenix, passing from the wilderness of Cattail Spring onto landscapes fragmented by freeways, fences, and suburban sprawl, doing so would entail a full reckoning with the reality of the American Southwest: with infrastructure that impedes migration, with the widespread persecution of predators, and most of all with livestock. More than 200 million acres of public land in the western United States are used for grazing cattle and sheep, an area half again the size of California, and on that land at any given moment can be found several million cows. Compared to them, the impacts of donkeys are a rounding error, but reducing livestock numbers is rarely considered. It's far easier to blame donkeys than to upset the status quo of industrial meat production.

Lundgren believes that donkeys are destructive only when human activity makes them so. Even then, some of that so-called destruction is a matter of perception. It's just how nature works. "If North American ecologists visited Africa more," he said—that is, visited a place where relatively intact populations of big animals still exist—"we'd understand that Earth's systems have always had areas that are trampled to shit and other areas that aren't." Diversity flourishes in the mix. The real problem is landscapes so transformed that they can't support basic, timeless ecological processes, and the reflexive categorization of donkeys as destructive actually normalizes a radically diminished modern state of nature. If anything, it's not donkeys who are not native. It's our own landscapes. An open-minded, case-by-case analysis of donkey impacts would make that impossible to ignore.

Were such an analysis conducted, perhaps it would be easier to confront all those other problems. And perhaps it would still be found that, in some places, their harms outweigh the values of their lives and their freedom, and that their removal is tragic but justifiable. Yet in those cases, management might be undertaken with a deep respect for the donkeys themselves. There might be renewed attention to sterilization programs, which are neglected and underfunded; attempts could be made to keep groups and families together. Conservationists might at least sit with the grief and moral residue of what is done. It might also prove that in some places, perhaps in many places, the presence of donkeys is indeed beneficial; that, rather than precluding life, they nourish more of it, and make their climate-changed, human-impacted desert habitats a richer and more resilient place.

And might these lessons apply to other so-called invasive species? Not every one will prove to be as beneficial as donkeys, or the historically South American apple snails upon whom endangered snail kites in Florida now depend. But some likely will. Certainly there will prove to be complexities obscured by a rush to judgment and revealed by closer examination—an examination motivated by a respect for every animal's life.

Special attention might be paid to animals in a similar predicament as donkeys, who are one of many large herbivores to flourish in regions outside their historical range. Wild horses are the obvious example, and there are 22 such species altogether: dromedary camels now found only in the deserts of Australia, hippos in Colombia, argali sheep in Western Europe, a whole host of large-bodied herbivores who are viewed as illegitimate by virtue of living where they once did not.

Many of them are, like donkeys, endangered or threatened in their historical homelands. Like donkeys, they are generally thought to be intrinsically destructive. The ecological roles for which they're celebrated elsewhere are ignored in their new homes. Often they're not even counted in official surveys of species populations and range. Dromedary camels are considered officially extinct in the wild, as though a brief detour into domestication should forever disqualify their lineage both from living freely and from membership in Earth's ecologies. Arian Wallach calls them "invisible megafauna," creatures erased "not from the world, but from our collective depiction of nature."

In a world where most of their large-bodied brethren are struggling—of 74 living large-bodied herbivore species, some three-fifths are threatened with extinction—these animals are surviving, even thriving, perhaps even restoring webs of life that would otherwise be lost. What—and who—might be seen if people looked upon those animals anew, with a sense of curiosity and wonder and care?

12

Wild Hearts

SOMETIMES A FEMALE WOOD FROG WILL LAY HER eggs in a puddle. It's an all-or-nothing proposition: on the one hand, there are few predators in a puddle compared to, say, a pond—no voracious dragonfly larvae or hungry bullfrogs—so her progeny have a better chance of surviving. On the other hand, that's only if the puddle doesn't dry out first. If it does, they all die. That would soon happen to these tadpoles: I had passed them on consecutive days, wriggling like so many spermatozoa in a parkside puddle that was smaller each day than the last.

I found an empty yogurt carton in a trash bin and scooped some up. Not far away was a drainage ditch that emptied into a creek. I released them there, went back to the puddle, collected some more, and repeated this for an hour or so until all were moved. Several hundred tadpoles doomed to bake in the sun now had a chance at life. I felt pretty good about it.

Was I being naive, though? Many of the tadpoles would soon die anyway. They'd fail to find enough food, be infested with parasites, or become food themselves. Even in the best of circumstances few tadpoles survive to adulthood. From that perspective I had merely exchanged their certain death for a near-certain, briefly postponed death. Even those who beat the odds would eventually end up in another animal's stomach—if, that is, they were lucky, the alternative being the slow death of injury or disease.

So maybe my heart should have been heavy as I delivered those tadpoles into the ditch down which they would swim into a world of suffering. Yet if that was all I thought about, all of nature would seem a nightmarish place. Instead, therefore, I might focus on the thriving of those who do survive; I could take solace in how these cycles make life's richness and abundance possible. Then again, that's easy to say when I'm not the one being eaten.

This reckoning likely crosses the mind of just about everyone at some point, and we find our ways of making peace with it. Some take comfort in chalking it all up to a divine plan; that's part of what originally made Darwin's theory of evolution so disturbing, as it implied a natural order of pitiless brutality and endless suffering untempered by the ministrations of God. Many precolonial cultures viewed animals as individuals but also as incarnations of a singular entity, and they believed that an animal's spirit could survive their earthly body's death. Some people attribute cruelty and even evil to predators, which at least provides the comfort of moral certitude.

The rise of an ecological ethic during the twentieth century—"He has not learned to think like a mountain," wrote Aldo Leopold of a rancher who exterminated wolves from his range—helped redeem predators. Environmental philosopher Holmes Rolston would go on to call predation a "sad good," tragic but necessary. This is where most nature-minded people have settled. Predators don't act maliciously; unconsciousness comes quickly; and predation, along with disease and starvation, is an essential part of nature. One might also speculate that the scientific denial of animal intelligence offered an indirect source of comfort: One doesn't need to grieve for an automaton.

But the advent of modern animal rights philosophy and the deeper scientific appreciation of animal minds that accompanied it brought renewed attention to the matter of animal suffering. To think that an animal's pains compare to our own and that their demise is the end of a precious, singular experience of life, makes it harder to shrug off misery and death as an inevitable necessity. Certainly we don't take such a circumspect attitude when it comes to our fellow humans or to animals we've come to love. Some philosophers have wondered if we might even do something about the suffering of wild creatures—or, at the least, better understand it.

They work in the field of welfare biology, a term coined in 1995 by the economist and philosopher Yew-Kwang Ng. Ng—who also helped lay the intellectual groundwork for effective altruism, a social movement devoted to using evidence and reason to determine the most effective philanthropic interventions—defined welfare biology as "the study of living things and their environment with respect to their welfare (defined as net happiness, or enjoyment minus suffering)."

That original paper was a rudimentary work. The science of animal well-being was still in its infancy, and Ng's calculations contained little ecology. Looking at it now is a bit like contemplating a mainframe computer from the 1970s: obviously a tremendous technical achievement but also woefully underpowered. Ng calculated that, on the whole, there was more suffering than enjoyment in the animal world—a conclusion he revisited in a 2019 paper coauthored with a PhD student who found an error in Ng's original model. In fact, "the model offers only ambiguity as to whether suffering or enjoyment predominates in nature," they wrote. The pendulum-swing scale of the discrepancy underscores how provisional these assessments are, but they are still signal achievements. However imperfect the conclusions, Ng was at least trying to fill a gap in human knowledge that, when you stop and think about it, is glaring.

If you want to understand, say, how a logging plan or change in rainfall will affect a particular forest, scientists can predict the impacts in myriad ways. Using computer models they can project, with extraordinary, report-filling precision, anticipated shifts in vegetation and species composition and all sorts of ecological and environmental metrics. Yet if you want to know what it all means for the day-to-day lived experience of the forest's residents—whether, on the whole, more rainfall means happier owls, or how that logging plan will affect the ratio of pleasure to discomfort in the lives of songbirds—then there are no models.

Trying to understand wild animals in this way is a natural extension of the fact that animals have a subjective experience of life, and the attempt has profound implications for how we understand nature. Does suffering truly predominate, casting a shadow over the whole wild world? And beyond that question, a finer-grained understanding of ecological interactions and the minds within them might help us better understand how ecosystems work—and even how to make them happier places for all.

◆ ◆ ◆

So is it fair to concentrate on life's thriving rather than its travails, and on those animals who survive rather than the fate of those who won't? "I think that's a way of choosing not to see the suffering," says Oscar Horta, a prominent figure among philosophers who study the issue of wild animal well-being. "But the suffering is not less real because we don't look at it."

Before Horta turned to philosophy, he studied economics. Peter Singer's now-classic *Animal Liberation* changed his path: Horta was already an animal person, and Singer translated those sympathies into rigorous intellectual frameworks of moral regard. Horta was mostly concerned with animals used for food and labor, though, and it was another treatise, Steve Sapontzis's *Morals, Reason, and Animals*, that turned him toward wild animals. Sapontzis was among the first ethicists to seriously consider the question of whether and when it was appropriate to help wild animals in need. An intervention might be wrong, he said, because it caused more harm than good—but he rejected the idea that intervening was wrong because suffering is natural. The natural order isn't automatically good.

In Horta's eyes, most people have an unrealistically idyllic view of nature. They think that, on the whole, good outweighs bad in the lives of wild animals, when in fact that's a romantic fiction to which people cling rather than confront uncomfortable truths. "What matters, really, is that there is suffering in the wild, and premature death," he told me. Sure, a few exceptional animals might have fairly decent lives—but "then there is this other thing, which is that some animals have lives that contain more suffering than pleasure."

Horta turned his attention to understanding the balance between positive and negative experiences in nature. Central to it is the fate of animals like the wood frogs I helped: species who have many offspring, most of whom die quickly. Traditionally known in biology vernacular as r-selected species, they're not the charismatic, long-lived beings who leap to mind when thinking about nature—the whales and elephants and so on—but they comprise the vast majority of wild lives. Their reproductive strategy is ideal for perpetuating their species, but far from ideal for any given individual.

Consider the results of a long-term study of wall lizards living in a cemetery in France: of 570 eggs, only 194 hatched and survived for a year, and by the fourth year, at which they would reach sexual maturity, a mere 12 lizards still lived. A wall lizard's maximum lifespan is a decade; premature death is their default experience— and wall lizards are reproductively frugal compared to Atlantic cod, an emblematic species in Horta's calculations, whose females lay several million eggs for every one who survives to maturity. For those extremely short-lived individuals, says Horta, the proportion of their lives spent suffering is immense. They don't live long enough for good experiences to accumulate on the other side of the ledger.

"The fact is that the overwhelming majority of the sentient beings who come to existence have to endure this fate," Horta wrote in an essay entitled, "Debunking the idyllic view of natural processes." As for so-called K-selected species, who have a few offspring of whom a comparatively large number survive, early death is still the norm. The fractions just aren't so extreme. For every adorable baby squirrel who survives to adulthood, several don't. Those individuals who do live long might well have relatively good lives—but even so they experience tremendous hardship. Unless they're apex predators themselves, they'll live in fear of being killed by a predator; they'll be afflicted by diseases and parasites, endure periods of famine and drought.

Altogether, argues Horta, there may well be far more suffering than well-being in the wild animal world. He acknowledges that more research is needed to be certain and to enumerate how the proportions vary by circumstance and species, but he is confident about the general conclusion. Many in the welfare biology community share it. And with that understanding comes a question: What should people *do* about it?

To Horta and others, it's unacceptable to shrug and say that suffering is an inevitable part of nature's cycles. After all, if human beings experienced similar levels of privation and premature death, we would be horrified. We would demand change. Nature's ends would not be seen as justifying its means.

This ethical imperative has led to some radical—albeit mostly hypothetical—ideas. To Horta, the notion that habitat loss is intrinsically bad is up for debate. It's terrible for those who lose their habitat, but it could prevent the suffering of future animals. "Given that most wild animals that are born have net-negative experiences, loss of wildlife habitat should in general be encouraged rather than opposed," writes the philosopher Brian Tomasik. Others argue that people should prevent wild animals from coming into existence and, when that's not possible, end their lives humanely; perhaps a few might be kept in a modern-day Noah's Ark where the worst suffering can be prevented.

Predation is prominent in these writings. In an essay published shortly after the controversial 2015 killing of an elderly lion named Cecil by a trophy hunter in Zimbabwe, Amanda and William MacCaskill that animal rights activists should *support* killing animals like Cecil. "Walter Palmer killed one animal," they wrote of the man who shot the beloved lion, "but in doing so he saved dozens of others." The philosopher Kyle Johansen shares their disapproval of predation but suggests a kinder alternative: perhaps someday it may be possible to use genetic engineering to create plants that provide sufficient nutrition to predators, obviating their need to kill.

Johansen also speculates about engineering r-selected species to produce fewer offspring. This is purely theoretical—even if the enormous technical challenges were surmounted, the likelihood of unintended, catastrophic consequences effectively renders it moot—but it still makes for a worthwhile thought experiment. And if, in that thought experiment, it turned out that those species couldn't function without having lots of offspring, perhaps they could be designed with pain-blocking neurochemistry so that their short lives were at least not quite so excruciating.

To be clear, such ideas are extremely far from mainstream among supporters of animal rights. Nearly every one of them I've ever asked about predation has replied that it's sad but necessary. They might intervene on behalf of a backyard squirrel threatened by a hawk, but they would also take that hawk to a wildlife hospital if the bird was injured. Many nature-loving readers likely had the same gut reaction to those suggestions about engineering nature as I did: a mixture of disbelief and revulsion. A world in which species evolved to produce many progeny have only a few, in which the ecological regulation provided by predation is absent, is a world of utter collapse. It's difficult to imagine it containing any animal life at all, or vegetation more complex than lichen and algae and perhaps a few wind-pollinated plants, although there might not even be soil—the product of r-selected life, of countless nematodes and worms and other invertebrates—for them to grow in.

Even so, these ideas are also fascinating. They represent an attempt to grapple head-on with the suffering of wild animals, and this dark view of nature is certainly not unique. "The whole system of nature," wrote the great biologist Alfred Russel Wallace, contemplating the implications of Darwin's theory of evolution, "has been founded upon destruction of life, on the daily and hourly slaughter of myriads of innocent and often beautiful living things." With properly attuned ears we would hear "sighs and groans of pain like those heard by Dante at the gate of Hell."

Richard Dawkins once wrote that "the total amount of suffering per year in the natural world is beyond all decent contemplation." During the minute it took to write that sentence, offered Dawkins, thousands of animals were being eaten alive—or perhaps from within, by parasites—or fleeing for their lives or simply dying from hunger, thirst, and disease. Should some seasonal plenitude ease those privations the relief would be only momentary, causing populations to swell "until the natural state of starvation and misery is restored."

And if nature is indeed so wholly suffused with suffering, to a degree that, if experienced by human beings, would make Earth a living hell—well, why would anyone with even a shred of empathy for animals *not* want nature to be radically reorganized and predators eliminated? Merely because we find nature beautiful, or attribute value to biodiversity or ecological integrity? Wouldn't those values be rendered awful in a world where they required near-infinite misery to be achieved? If the argument in nature's favor would then be that humans rely on natural processes to survive, on vegetation to produce oxygen, and on insects to pollinate our crops, wouldn't the preservation of nature be only a matter of naked self-interest, an original sin at the heart of human existence?

This line of thinking is terrible indeed. It is the most troubling implication of red-in-tooth-and-claw welfare biology, and far more important than far-fetched scenarios of biological engineering. It's a modern gloss on attitudes that for centuries have motivated—or at least helped to rationalize—the destruction of nature and persecution of wild creatures. There's little chance that wood frogs will ever be designed to lay a precious few eggs and attend them with the care of a mother elephant for her calf; but it seems all too possible that people could come to see nature through a dark lens, as not worth caring about, much less fighting for.

◆　◆　◆

NOT EVERYONE THINKS THE STATE OF NATURE IS QUITE so overwhelmingly miserable. "My worry is that in trying to push back against this," says philosopher Heather Browning of the welfare biology community's rebuke of sanitized, idyllic notions of animal life, "perhaps we've swung too far in the opposite direction." Perhaps they have over-emphasized suffering and made it seem more dominant than it actually is.

Browning grew up in Australia, and after finishing college took a job as a keeper at the National Zoo and Aquarium in Canberra. (Among her favorites, says Browning, were the dingoes, whom she took for morning strolls before the zoo opened.) Eventually she became an animal welfare officer; the job led her to reflect on the very essence of her task. How do people decide what good—or bad—welfare is? How do they measure it? What assumptions do they bring with them? She went back to school, earning a PhD in animal welfare science before joining the London School of Economics' Foundations of Animal Sentience project.

According to Browning, some of what people consider to be terrible animal suffering probably isn't quite so bad. Predation makes for gruesome descriptions of animals being chased down and ripped apart, but it also tends to happen quickly. When a wildebeest's life ends in a lion's jaws, suffering is intense but brief, occupying only a few minutes in a life that lasted for years. There's also reason to think that many animals benefit from what reptile biologist Clifford Warwick calls "incidental compassion." He gives the example of how, when shutting one's finger in a door, pain comes immediately, yet people who lose limbs in accidents often report feeling nothing until much later. The body's own shock-induced biochemistry dulls the pain, which makes good evolutionary sense: pain is useful in imparting lessons about what not to do, but being overwhelmed by pain at moments of extreme duress is unhelpful.

It is also the case that some immature animals are cognitively much simpler than adults. Their brains and nervous systems are incompletely developed; not yet fully conscious, they have what Browning calls reduced sentience, and the experience of pain may be correspondingly stunted. Perhaps those tadpoles would not have suffered to the degree that I thought. This is still speculative, cautions Browning, and even if true will likely vary depending on the biology of a given species. But it at least offers the possibility of respite.

Yet even if that's not the case, and immature animals suffer as much as adults do, does it necessarily mean that suffering outweighs pleasure in the animal world? Browning has her own take on Horta's cod example. Yes, all those billions of juvenile cod do suffer, but add up all the minutes and seconds of their pain and it's outweighed by the moments of good experienced by cod who survive to adulthood. Browning takes issue with welfare biology's emphasis on pain. Predation, drought, floods, extreme cold and heat, disease, parasites, competition: all these are unquestionably a part of animal existence, she acknowledges, but what about the good in their lives?

She cites Simona Ginsburg and Eva Jablinka's *The Evolution of the Sensitive Soul*, a treatise on the evolution of consciousness, in which Ginsburg and Jablinka float the idea that life has an intrinsically positive valence. Simply existing feels slightly good. This is the great innovation of consciousness: it motivates an organism to keep going. Atop that foundation are further layers of simple, fundamental pleasures: exploration, paying attention, learning, looking, moving, exercising agency. The necessities of life, each with an emotional incentive. One of Browning's favorite examples is how birds' brains release opioids as they sing. "These are things that we're only just beginning to explore in the world of animal welfare," says Browning. "There seems to be so many more sources of pleasure in the world than what we thought."

Browning and Horta both agree that more research is needed on the well-being and interior lives of wild animals. Scattered pieces of such research already exist, and the intellectual scaffolding is in place for something more comprehensive. On my computer is a folder of studies by conservation biologists and other scientists who measure biomarkers—such as the production of stress hormones or cellular changes linked to stress and premature aging—to gain insight into wild animal welfare. This research, to pick a few representative examples, informs us that chipmunks in urban areas may actually have easier lives than those in intact forests, but amphibians tend to find urbanization especially difficult; that carrion crows with stronger social bonds also tend to have better physical health; and that early life is especially stressful for Seychelles warblers, but gets much easier once they've found a territory and established their status.

These studies make fairly straightforward connections between stress and physical well-being. Other research has started to probe subtler questions. One experiment involving captive Sumatran orangutans, for example, offered them two paths, both of which ended with a food reward, but only one of which involved making a decision about where to go along the way. The orangutans consistently took the option that involved a choice, as do humans when presented with similar tests. Being able to make choices—to engage with one's environment and exercise agency—is intrinsically rewarding to both our species. Another experiment involving zebrafish described how investigating an unfamiliar place lifted their moods, hinting at how widespread that satisfaction may be among animals.

Such findings only scratch the surface of what might be studied. Scientists are starting to think about how to measure the subjectivities of boredom and happiness. Perhaps they might someday engage with the profound philosophical question of what it means for an animal to lead a meaningful life. Might a bear understand that berry bushes grow from their scats, a beaver feel good about their dam's solidity, or a phoebe look with satisfaction upon a well-built nest? What is the mental experience of those carrion crows for whom social connections improve physical health? How important is affection to them—or, for that matter, to a lake sturgeon? (Lake sturgeon, for the record, are a social fish, and are less stressed by extreme heat when they experience it in the company of other sturgeon than when they're alone.) And what might be vital to animal well-being, to the richness of their lives, that is outside the present range of our imaginations, much as the ability to navigate by Earth's magnetic fields is beyond our physiological pale?

Meanwhile the science of welfare in captive animals is already fairly well-established. The mainstream paradigm is called the Five Domains model, which assesses animals according to their nutrition, environment, health, behavior, and mental state. Applying this to wild animals is a technical rather than existential challenge. Researchers led by David Mellor and Ngaio Beausoleil, both of them animal welfare specialists in the veterinary school at New Zealand's Massey University, recently published a protocol for assessing the welfare of wild animals; they demonstrated it with wild horses, but it could be adapted to other species. Mellor, Beausoleil, and colleagues propose the creation of a new discipline, "conservation welfare," that would expand conservation biology's focus on animal fitness to include their feelings as well.

Such research remains esoteric, but it doesn't need to be. There is even an organization devoted to studying wild animal welfare: Wild Animal Initiative, founded in 2019. "It's all toward building a new academic field," said Cameron Meyer-Shorb, the organization's executive director, with more than a little millennial startup cockiness. (Both he and strategy director Mal Graham are in their early thirties.) One of their first research projects, led by biologist Luke Hecht, involved trying to estimate individual well-being according to population dynamics—in other words, how likely is a wood frog egg to hatch into a tadpole? To grow legs, become a froglet, and finally develop into an adult? What is their typical experience of life at each stage, their balance between pain and pleasure, and how does all that translate into the average expected welfare of a typical frog?

Other projects are narrower in scope. In 2022 and 2023 they disbursed $1.2 million in grants—not a huge amount compared to what government science-funding distributes, but nothing to sneeze at, either—for dozens of projects, including research on salamander welfare and their experience of environmental change; measuring farming's impacts on the well-being of wild caterpillars; studying the emotional lives of octopuses; using thermal imaging to measure early-life stress in birds; and assessing how the welfare of juvenile Murray cod is affected by water quality.

Meyer-Shorb's enthusiasm notwithstanding, this type of research isn't entirely new. Some of it already takes place, in a less concentrated fashion, under the umbrella of conservation biology and ecology. One can imagine—in an ideal world, anyway, where there's no shortage of resources for studying every question of interest and importance—studies like these being conducted for every species, in every ecosystem.

Then things would get *really* complicated: How does well-being play out not only at the level of species and populations considered in isolation, but also when species and populations interact with one another and their environments in vast tangled webs of ecologies? What does the presence of a whale mean for the well-being of all the aquatic creatures whom their life touches? To understand these questions is a daunting, even overwhelming task—yet if we can plot the movement of atomic particles in the moments after the universe formed, surely we can do more than speculate about whether a sparrow outside the window is happy.

◆　◆　◆

EACH MAY I VISIT LEONARD'S MILLS, A RECONSTRUCTED eighteenth century logging camp beside Blackman Stream in Bradley, Maine. Blackman Stream is a tributary of the Penobscot River, the second-largest river in New England, where over the last decade several large hydroelectric dams have been removed. Millions of alewives and blueback herring poured upriver and into streams like Blackman, hearkening back to a time when, in the words of Thoreau, New England's inland waters ran silver with the bodies of migratory fish returning from the ocean.

As they leave the Penobscot and course up its tributaries, however, fish still find their way blocked by small dams that were built centuries ago to power streamside mills. At least at Leonard's Mills there's a fish ladder, a series of artfully cut stone pools that fish may follow around the dam—but some don't find it. Instead they conclude a thousand-mile migration by hurling themselves into the dam's roaring runoff-fed spillover, again and again and again, until their energies are spent. Then they rest in the slower water at the stream's edge, so exhausted that I can pick their muscular, journey-scarred bodies up with my hands, drop them into a bucket, and carry them to the pond above.

Several times each May I spend an hour or so doing this. It seems only right, although for some reason I don't feel quite the same gut-level urgency I did when rescuing the wood frog tadpoles.

Certainly it doesn't occur to me to wonder whether their lives contain more pain than pleasure. Perhaps that's because I don't empathize so readily with fish, but I suspect it also has something to do with how links between the fishes' perishing and the surrounding landscape's flourishing are so evident.

At this time of year trees have only recently unfurled their leaves but flowers on the forest floor, now shaded but having drunk deeply on early spring sunshine, are in bloom. Along the path to the dam are colonies of gaywings, *Polygala paucifolia,* one of my favorite spring ephemerals, with tiny purple-pink blossoms that look like balloon-animal dogs. Beside the stream is a profusion of ferns and shrubs, and above the shadbushes—named for how they flower at the same time as shad, another migratory fish, return—swallowtail butterflies catch sunbeams. Eastern phoebes returned from their own migration perch in the branches, darting down to catch mayflies. All this life is nourished, directly or indirectly, by the bodies of the river herring: So many fish gather at the ladder and below the dam that they can be smelled from hundreds of feet away, and along the pools beneath the dam are the remains of herring eaten by mink and raccoon and all the other creatures who feast here at night. Their bodies become the soil; they become

the flowers and ferns and swallowtails. The Passamaquoddy people call alewives "the fish that feeds all."

The forest surrounding Leonard's Mills is owned by the University of Maine and the US Forest Service, whose researchers use it to study forestry and ecology, and near the pond are study sites belonging to Alessio Mortelliti, an ecologist. He too is interested in animal intelligence and the lived experience of forest creatures, but for a different reason than the welfare biology philosophers. Rather than wondering whether their lives are bad or good, he wants to know how their mental properties fit into the grand tapestry of ecosystem function. Mortelliti studies small rodents, primarily voles and mice, and their harvesting of seeds: a seemingly tiny process, but up to 95 percent of the cones, seeds, and nuts in a given patch of forest are collected by rodents, and those they don't eat become the forest's next generation.

For the past several years Mortelliti and his students have spent much of their time catching small rodents—roughly 5,000 in all—and, before tagging and releasing them, putting them through a series of tests designed to measure whether they are shy or bold. Come autumn they watch those rodents gather seeds: what kind they choose, whether they eat them immediately or hide them, and whether, if they hide them, they find them again. Among their tools are trays of seeds surrounded by a nontoxic fluorescent powder that temporarily adheres to the rodents' feet and leaves behind a trail. One predawn October morning I accompanied Mortelliti's collaborator, Allison Brehm, as she inspected the night's activity. Under her ultraviolet lamp the forest was transformed, its mossy floor and fallen trees overlaid with constellations of tiny footprints that led Brehm to each seed's fate. The sight made me see the forest anew. It did not just spring into being. Rodents planted it.

The research has yielded a uniquely detailed map of the forest: the territories of thousands of individual voles and mice, their personalities, vegetation structure, the presence of predators, and how all these aspects interact with one another. Where a forest has been logged so that its trees are homogeneous in age and composition, rodent personalities are homogeneous too; when a forest is left intact, or managed to resemble a natural forest, both shy and bold rodents are found. Although both will eat the seeds of trees they haven't before encountered, the latter are more likely to carry them across longer distances—something that may help trees shift their ranges as the climate changes. Shy rodents are most likely to cache seeds in fallen logs, which are optimal sites for germination. In certain circumstances the rodents are said to have a mutualistic relationship with particular species, planting enough of their seeds that they help the trees proliferate. At other times they're considered predators of those species, consuming so many seeds that the trees' presence declines.

Mortelliti is careful to say that these processes are nuanced and much remains to be learned, but his findings point toward an overarching principle: Having a diversity of rodent personality types makes a forest richer and more resilient. Depending on the vagaries of drought, fire, and predator fluctuations, different personality types may come to the fore. One could even imagine how, in a world where personality did not yet exist in small seed-dispersing animals, its evolution would help the assemblages of vegetation we call forests to regenerate themselves, in turn making them more hospitable to those animals—and others too, of course—in a virtuous cycle of cognition and verdancy.

Through that lens the evolution of personality might be understood as a function of ecology. Minds don't only serve the individual animal, but their community, too. To our understanding of fundamental forest processes, of worms digging in earth and fungi spreading through fallen wood and millipede legs churning leaf litter to create the richness of soil, infinitesimal actions repeated at planetary scales, one can add the decisions of mice with seeds in their cheeks.

Integral to these processes is predation. On average, mice in a Maine forest don't live to see their first birthday: They're eaten by foxes, skunks, weasels, mink, bobcat, coyotes, hawks, and owls, whose presence creates what ecologists call a "landscape of fear." Within it rodent behaviors—how far, for example, a red-backed vole will stagger vulnerably beneath the weight of an acorn before finding a hiding spot—are shaped by the threat of predation. Take that threat away and the resilience of a forest is jeopardized. Yet does that mean, per the concerns of the welfare biologists, that one could understand the forest's health as a product of unrelenting fear?

I asked Brehm what she thinks of this. She is a kind-hearted person, not given to ignoring the rodents' value as individuals; the previous winter she cared for a baby white-footed mouse whom she found in her house, keeping the mouse safe and well-fed until springtime, and she does not doubt that mice have their own meaningful experience of life. She suggested that the so-called landscape of fear might better be understood as an ecology of carefulness. Mice don't live in constant fear but instead perceive risk and act accordingly.

Evolutionary theory suggests this may be a better way of understanding their experiences. Long-term, chronic stress has all sorts of unpleasant physical effects in animals, from depressed immune systems and metabolic disruption to shortened lives; constant fear would be maladaptive. Studies suggest as much. In Australia, where dingoes suppress fox populations, foxes do avoid dingoes, but are no more cautious or vigilant when their territories overlap. They actually seem to be *more* active and confident; they simply know when dingoes are active and avoid them. The key is social stability. So long as dingoes are not being killed—unfortunately dingoes are persecuted much as coyotes are in North America—foxes can learn the habits and predict the behaviors of their larger neighbors.

Similar findings have been made in zebras who, in an area with lions, spent no more time being vigilant and produced no more stress hormones than in areas without lions. A study of seals living on islands off the coast of South Africa found that, so long as they could respond to the risk of shark attacks by hiding in kelp beds, their stress levels were no higher than in the absence of predation. The researchers responsible for the fox and dingo study suggested that animals have cognitive maps of their homes. They called this a landscape of knowledge. Rather than fear, they are guided by risk evaluation.

That offers the possibility that fear is less ever-present than one might think. And without predators, populations of seed-dispersing rodents would fluctuate even more than they already do. There would be more years when, as rodent numbers rose and seed abundance ebbed, hardly any would be planted. Predation enables flourishing. That does not make the act of killing easy to contemplate; it doesn't bar us from intervening on behalf of a wild animal for whom we care, just as I shoo foxes away from the groundhogs who summer in a burrow beneath my deck. Death is still sorrowful and each lost life should be acknowledged—but, writes bioethicist Bill Lynn, "It is not pain and suffering per se that is wrong, but unnatural pain and suffering."

As for the river herring flowing up Blackman Stream: Brehm observed that the bodies of those who don't make it, who get stuck below the dam or in a series of difficult-to-navigate rocks, nourish predators and scavengers at an especially slim time of year when winter has receded but summer's bounty is yet to arrive. I am sorry for those fish and also grateful for them. Rescuing a few is an expression of care for individuals and gratitude for the population, but I am not trying to refute death.

◆　◆　◆

MORTELLITI'S RESEARCH ON RODENT PERSONALITY and forest regeneration offers a hint of what might be learned. Perhaps a time will come when we understand how the memories of hermit crabs shape coastal ecosystems and the problem-solving abilities of elephants influence savannah fire cycles. In the meantime, the research raises an interesting possibility: that when managing nature, people might consider the minds of animals. In the woods of the northeastern United States, say Mortelliti and Brehm, foresters could adopt harvesting practices that mimic natural conditions, preserving a diversity of tree species and life stages that in turn preserves a diversity of rodent personalities.

That approach isn't explicitly about improving the lives of animals, although it's possible that, all things being equal, rodents in more natural forests tend to have a better experience of life. But other interventions are more direct. Immunization programs for wild animals already exist, albeit usually for the purpose of preventing diseases from spilling over into humans, as with rabies vaccinations for raccoons. These might become more common and performed for animals' own sake. Perhaps wild birds could be vaccinated against virulent new strains of avian influenza. People can try to assist animals affected by extreme weather—buckets of water for animals are already a common suburban feature during drought and wildfire seasons in western North American cities—or even help move them to more hospitable places, as some biologists have suggested for alpine pikas stranded by climate change on mountains that can no longer support them.

Some of those ideas overlap with what conservationists already do, although their work isn't usually described in terms of animal happiness. With a better understanding of how ecology and animal experience intertwine, such that welfare can be predicted and measured as precisely as vegetation change or nutrient cycling in different systems and contexts, people might even think about how to manage ecosystems to enhance well-being. Oscar Horta speculates on how people could eventually nudge ecosystems toward configurations that favor species with high lifetime well-being; perhaps people will find that particular types of vegetation or habitat features have outsized effects on animal happiness. (Someday I would love to see a study on the cumulative experiences engendered by big old hollow trees.) Cat Kerr of the Wild Animal Initiative suggests that ecosystem rewilding and restoration projects "offer the opportunity to be really intentional" and "design these habits to maximize welfare."

Some people may find it unusual, even inappropriate, to intervene in nature for that purpose. Yet humans already intervene constantly in natural processes in order to achieve environmental outcomes: to increase biodiversity, restore ecosystem function, and protect species, not to mention the far more radical and routine interventions of housing, farming, forestry, and mining. If we're already shaping a habitat, why not consider animal well-being as an outcome? "The question is not 'Should we intervene in nature?'" writes Horta, "but rather, 'in what ways should we intervene?'"

That level of understanding is still far away, though. In the meantime I'm content with some simple rules of thumb: that healthy ecosystems mean happier animals, and that we ought to restore and protect them. I once saw a presentation about people building piles of stones in a desert wash through which water drained, violently and briefly, after rainfall. The rocks would slow the water, allowing it to enter the ground and collect in standing pools, eventually forming a wetland. Beavers had already returned and were adding their dams to the effort. As ecosystem restorations go this was not an especially remarkable project, but it was a beautiful one, and the sheer amount of labor involved was poignant. To make a happier world, people carried stones through a desert.

On this, both animal advocates and traditional conservationists might find common ground. Perhaps this could draw people who are passionate about animals toward conservation. At present our society has two great movements dedicated to other-than-human life—yet they keep each other, for the most part, at arm's length. And conservationists might understand the protection of habitat, wilderness, wildness, and nature, through a slightly different lens than usual: as nourishing not only biodiversity, but the happiness and aspirations of our nonhuman neighbors.

Conclusion

AS THE FIRST RIVER HERRING GATHER AT LEONARD'S
Mills, the amphibian migrations are coming to an end: a quiet
flood of springtime life, less renowned than the migrations of
birds and even fish, perhaps because they occur on rainy nights.
It is then that amphibians come, leaving the leaf piles, hollow logs,
and old burrows where they've spent the winter, walking, hopping, and wriggling across distances that our striding legs would
cover in a few minutes but are, at their scale, an epic pilgrimage.
Like the river herring they return to their birthplaces—not rivers
and streams but instead vernal pools, the forest depressions that
fill with water each spring and become fountains of woodland life.

Most numerous are spring peepers, fingertip-sized frogs loud
enough in chorus to drown a jet engine. In my corner of Maine
there are also wood frogs aplenty, their duck-like voices always
coming as a slight surprise, and leopard and pickerel frogs with
mottled skin as beautiful as their namesakes. They can live for

nearly a decade, which seems a small miracle. How can something so small and soft, so vulnerable, persist for so long?

Their lifespans are short when compared to the salamanders who are the luminaries of these magical nights. There are eastern newts with bright orange skin suggestive of the fires in which medieval Europeans, perhaps not realizing that the creatures hibernated in their woodpiles, believed they were born; they can live for 15 years. Spotted salamanders—whose embryonic days are spent in symbiosis with algae, allowing them to be, at least for a while, photosynthetic—can live for 30 years. Blue-spotted salamanders may live even longer. Some belong to a subpopulation, identifiable only by their DNA, composed entirely of females who reproduce by cloning themselves, which seems appropriate to their understated marvelousness.

The salamanders eat small invertebrates, including slugs, earthworms, and millipedes, who eat decomposing vegetation: the very foundation of the trophic pyramid. This foundation goes unnoticed by most, but in a given patch of forest the salamanders' biomass may outweigh that of birds and small mammals. Yet unless one is given to looking beneath rotting logs, it's easy to go a lifetime without seeing a salamander. I saw my first only a few years ago. Their journeys, after all, are hidden in darkness—and what journeys they are! When scientists plotted wood frog migration distance as a function of body weight, their quarter-mile perambulations eclipsed the many-hundred-mile migrations of wildebeest. The same could be said of other amphibians. In a world latticed by roads, those migrations all too often end beneath an automobile's tires—a tragic fate for any animal, but magnified by the beauty of these animals, their strangeness, their softness.

Several years ago I heard about a young man named Greg LeClair who as a child had gone out on those rainy spring nights to carry amphibians across roads. A few, at least, would be escorted to safety in his hands. As a high school student he founded a Facebook group to coordinate the efforts of other volunteers. Later he

added a community science component, with the rescuers counting those they saved—and those they didn't—so that, when road
maintenance time came, amphibian crossings might be installed
in the most dangerous places. I signed up and now spend rainy
April nights walking back and forth on a stretch of road outside
Bangor, clad in a safety vest and headlamp, trying to reach the
amphibians before they're run over.

Even moderate traffic at the wrong time can wipe out a population in a few years, but I'm not out there in the rain solely for a population's sake, or to do my part to prevent extinction, or because
amphibians are an important link in food chains and distribute
nutrients across landscapes and enhance the diversity and abundance of life they contain. All that is part of the equation, but what
completes it are the individual animals: that moment of connection with a frog or a salamander.

Thanks to a quirk of amphibian physiology their mouths are set
in what looks like a grin. Of course it doesn't mean they're actually smiling—although, as their vernal pool sojourns culminate in
several nights of frenzied communal mating, maybe they are smiling a bit—but it's a useful reminder that, whatever they're feeling at the moment, they're feeling *something*. That, as different as
they are from me, they're still someone, living in the first-person,
beings with an experience of life as singular and rich as the veined
golden ring around a pickerel frog's eye.

All those calculations of suffering and well-being, of the difficulties their lives will contain, how many of the eggs they lay
will be consumed, and how nearly all their progeny won't survive
to adulthood: None of it matters in the end beside the conviction
that they should not die tonight, beneath the wheels of an automobile, and that each one I can save is a flicker of light in this world,
a grain of sand on the scales of life. And so on these rainy spring
nights, the air rich with the smell of earth, of decomposition and
stirring, so welcome after the long winter, I walk back and forth
along the road.

When my headlamp illuminates the telltale bulge of a salamander or a frog, I trot toward them, scooping the animal up in my hands and carrying them across the road, in whatever direction they were going, before setting them down in the brush. Most of the time I manage to reach them before a car does, but there are always those who cross at the wrong time, or—cruelest of all—those I see in the headlights of an oncoming car, too late to do anything except pick up their mangled bodies afterward and move them to the verge so that at least scavengers won't be struck, too.

Sometimes I think of Amanda Stronza, a conservation anthropologist who several years ago started posting photographs to Facebook of the memorials she made for animals killed by cars. She would carry them off the road and into nearby woods or brush, then surround their bodies with radiating circles of flowers, leaves, and grasses: something to honor their singular, precious lives, something more dignified than lying bloated on the tarmac. The photos went unexpectedly viral.

"What has surprised me," Stronza told me, "is how many people have said, 'Oh, I do this too,' or 'I've always thought about doing this.' It makes me feel like there's a really big community out there, the people who want to relate to animals in an intimate way—but it's almost like they're afraid people will laugh at them." There is so much care out there, so much consideration, just waiting for permission.

On these nights there isn't time for ceremony, though. Every moment delayed increases the likelihood of being a few steps too late for someone who could still be helped. On the first spring I did this, the COVID pandemic had recently erupted and there was hardly any traffic at all; now it has returned, a car every minute or two, more if rains fall early in the evening or on a weekend night. They roar past, headlights burning spots in my retinas so that it becomes harder to see the amphibians, and at those times the exercise feels like it's about more than a road in eastern Maine. It's a metaphor for a time of mass extinction, for a world full of life being obliterated by thoughtlessness, greed, and a doctrine of endless economic growth, the costs of which are borne by those who cannot object. A world of beauty chewed up and spit out, not because it's necessary but because the people who would change it are eclipsed by those who choose not to. A world in which a decent future seems ever more improbable. On nights like the loss is made flesh.

Will an understanding of other animals as thinking, feeling individuals, an ethos animated by empathy and consideration, turn that terrible tide? On its own, probably not—just as traditional conservation and environmentalism have not sufficed on their own, either. If humanity is to be more than a biological asteroid, then all these ways of thinking are necessary. Together we have a chance.

To care about others is to be constantly reminded of how much pain and struggle there is in the world, and also how it is within each of our grasps to make a difference that is, for someone, world changing. There is so much that can be done, collectively and individually, to live better—more kindly, more thoughtfully, more fairly—with the animals around us. *Meet the Neighbors* is by no means exhaustive; the point was to feed your imagination, to encourage you to think about what else is possible and within your grasp.

If the crises of climate change and biodiversity collapse are not averted, then these ideas will be no less important. We will still need to nourish islands of care and abundance and to make the world better for those who survive. For our *neighbors*.

What is there to do, except what one can? Far down the road I see the glimmers of other headlamps. In the years since I started, the membership of Maine Big Night has swelled from a few hundred to several thousand people, all of them out there in the darkness, saving lives.

The cars roll by, and in between them I gather as many frogs and salamanders as I can. And there is another salamander, so fragile and vulnerable, silhouetted by an oncoming truck.

For a moment I think I can dash out but it's coming too fast. My heart sinks; after a while one has a sense of when an amphibian might pass safely beneath the wheels and when they won't, and this one is right in the path. The truck passes in a roar of tire and engine. Out I trudge to pick up the body. There is the salamander, unharmed.

Acknowledgments

IT TAKES A VILLAGE TO WRITE A BOOK, AND THE mayor of *Meet the Neighbors* is my editor, Jessica Yao, whose thoughtfulness, vision, and care—and patience!—made this work immeasurably better.

Thank you to all the scientists, scholars, activists, and practitioners whose labors laid the intellectual foundations of *Meet the Neighbors*. A sentence or two often encapsulates years of their toil; the efforts underlying these pages are truly vast.

I am indebted to everyone who appears in them for their expertise and generosity, but a few people went above and beyond duty's call: Megan Adams, Paula Goldberg, Janet Kessler, Erick Lundgren, Kathy Pollard, and Steve Wise. Many people were not quoted but provided invaluable insight, and I am grateful to them as well: Kyle Artelle, Rachel Atcheson, Chelsea Batavia, Jonaki Bhattacharyya, Sumath Bindumadhav, Kevin Bixby, Jabari Brisport, Kendra Coulter, Nancy Ganner, John Griffin, John Hadidian, William Housty, David Karopkin, Grace Kuhn, Robb Lansdowne, Barry

MacKay, Aaron Mills, Amber Prince, Lauren Riters, Allen Rutberg, Elizabeth Stein, Tracy Timmins, Zoe Todd, Maria Condoy Truyenque, Giorgio Vallortigara, and Arian Wallach.

Thank you to the Creature Club—Rebecca Giggs, Ben Goldfarb, Holly Haworth, Ferris Jabr, and Rob Moor—for their chapter readings and advice. I am also thankful to everyone on the Slackline, and especially Laura Poppick, for supporting me through the arduous task of writing. Andrew Westoll's counsel to remember the joy helped me push through the final stretch.

Thanks to my agent, Jill Marsal; Quynh Do; and everyone at W.W. Norton, including Annabel Brazaitis and copyeditor extraordinaire Marjorie Anderson, who helped bring *Meet the Neighbors* into the world. And a special thanks to my mother, Angela Gilladoga, whose discipline and compassion is a perpetual inspiration.

Last but absolutely not least, I thank all those creatures whose presence makes this world an infinitely richer place.

Notes

INTRODUCTION

000 **Fiona Presly spotted a bumblebee:** Unless otherwise noted, information about Fiona Presly and Bee comes from interviews with the author.

000 **A video on *The Dodo*:** "Woman Becomes Best Friends with a Bee She Rescued." Soulmates: An Original Video Series by *The Dodo*. March 21, 2018.

000 **skyscraper-climbing raccoon:** Keim, Brandon. "It's Not About the Raccoon." *The Atlantic*, June 15, 2018.

000 **goose looking through the door:** Shimano, Mihiro. "An Unexpected Visitor: Wild Canada Goose Turns Up at Animal Hospital to Await Mate in Surgery." *Boston Globe*, July 16, 2021.

000 **most ... are dwindling:** Leung, Brian, Anna L. Hargreaves, Dan A. Greenberg, Brian McGill, Maria Dornelas, and Robin Freeman. "Reply to: Emphasizing Declining Populations in the Living Planet Report." *Nature* 601, no. 7894 (January 27, 2022): E25–26. https://doi.org/10.1038/s41586-021-04166-y.

000 **learn by observation:** Bridges, Alice D., HaDi MaBouDi, Olga Procenko, Charlotte Lockwood, Yaseen Mohammed, Amelia Kowalewska, José Eric Romero González, Joseph L. Woodgate, and Lars Chittka. "Bumblebees Acquire Alternative Puzzle-Box Solutions via Social Learning." *PLOS Biology* 21, no. 3 (March 7, 2023): e3002019. https://doi.org/10.1371/journal.pbio.3002019.

000 **and by association:** Russell, Avery L, and Tia-Lynn Ashman. "Associative Learning of Flowers by Generalist Bumble Bees Can Be Mediated by Microbes on the Petals." *Behavioral Ecology* 30, no. 3 (June 13, 2019): 746–55. https://doi.org/10.1093/beheco/arz011.

000 **count:** Skorupski, Peter, HaDi MaBouDi, Hiruni Samadi Galpayage Dona, and Lars Chittka. "Counting Insects." *Philosophical Transactions of the Royal Society B: Biological Sciences* 373, no. 1740 (February 19, 2018): 20160513. https://doi.org/10.1098/rstb.2016.0513.

000 **recognize by sight ... vice versa:** Solvi, Cwyn, Selene Gutierrez Al-Khudhairy, and Lars Chittka. "Bumble Bees Display Cross-Modal Object Recognition between Visual and Tactile Senses." *Science* 367, no. 6480 (February 21, 2020): 910–12. https://doi.org/10.1126/science.aay8064.

000 **using tools:** Loukola, Olli J., Cwyn Solvi, Louie Coscos, and Lars Chittka. "Bumblebees Show Cognitive Flexibility by Improving on an Observed Complex Behav-

ior." *Science* 355, no. 6327 (February 24, 2017): 833–36. https://doi.org/10.1126/science.aag2360.

000 **have feelings:** Solvi, Cwyn, Luigi Baciadonna, and Lars Chittka. "Unexpected Rewards Induce Dopamine-Dependent Positive Emotion-like State Changes in Bumblebees." *Science* 353, no. 6307 (September 30, 2016): 1529–31. https://doi.org/10.1126/science.aaf4454.

000 **a commentary on Bee:** Chittka, Lars. "A Bee as Pet—A Bee Psychologist's Perspective." *Antenna* 42, no. 1 (2018): 4–5.

000 **600 million years ago:** Zimmer, Carl. "To Bee." *National Geographic*. October 25, 2006.

000 **snake friendships:** Skinner, Morgan, and Noam Miller. "Aggregation and Social Interaction in Garter Snakes (*Thamnophis sirtalis sirtalis*)." *Behavioral Ecology and Sociobiology* 74, no. 5 (May 2020): 51. https://doi.org/10.1007/s00265-020-2827-0.

000 **songbird syntax:** Suzuki, Toshitaka N., David Wheatcroft, and Michael Griesser. "Experimental Evidence for Compositional Syntax in Bird Calls." *Nature Communications* 7, no. 1 (March 8, 2016): 10986. https://doi.org/10.1038/ncomms10986.

000 **bison deliberations:** Ramos, Amandine, Odile Petit, Patrice Longour, Cristian Pasquaretta, and Cédric Sueur. "Collective Decision Making during Group Movements in European Bison, *Bison bonasus*." *Animal Behaviour* 109 (November 2015): 149–60. https://doi.org/10.1016/j.anbehav.2015.08.016.

000 **puzzle-solving turtles:** Davis, Karen M., and Gordon M. Burghardt. "Long-Term Retention of Visual Tasks by Two Species of Emydid Turtles, *Pseudemys nelsoni* and *Trachemys scripta*." *Journal of Comparative Psychology* 126, no. 3 (August 2012): 213–23. https://doi.org/10.1037/a0027827.

000 **self-aware fish:** Kohda, Masanori, Redouan Bshary, Naoki Kubo, Satoshi Awata, Will Sowersby, Kento Kawasaka, Taiga Kobayashi, and Shumpei Sogawa. "Cleaner Fish Recognize Self in a Mirror via Self-Face Recognition like Humans." *Proceedings of the National Academy of Sciences* 120, no. 7 (February 14, 2023): e2208420120. https://doi.org/10.1073/pnas.2208420120.

000 **rats who feel regret:** Steiner, Adam P., and A. David Redish. "Behavioral and Neurophysiological Correlates of Regret in Rat Decision-Making on a Neuroeconomic Task." *Nature Neuroscience* 17, no. 7 (July 2014): 995–1002. https://doi.org/10.1038/nn.3740.

000 **a "new Copernican revolution":** Chittka, Lars. "The Intelligent Mind of an Insect." *Interalia Magazine*, November 15, 2022. https://www.interaliamag.org/articles/lars-chittka-the-intelligent-mind-of-an-insect/.

000 **"We are as gods . . . .":** Daloz, Kate. *We Are as Gods: Back to the Land in the 1970s on the Quest for a New America*. (New York: PublicAffairs, 2016), xii.

000 **as . . . Aldo Leopold put it:** Leopold, Aldo. *A Sand County Almanac, and Sketches Here and There*. (Oxford, UK: Oxford University Press, 1989), 204.

1. THE ESSENCE OF INTELLIGENCE

000 **His final book:** Darwin, Charles. *The Formation of Vegetable Mould Through the Action of Worms: With Observations on Their Habits*. (London: John Murray, 1881).

000 **Feeling only pain or hunger:** Aristotle, *On the Soul*. J. A. Smith, trans. (Stiwell, KS: Digireads.com, 2006).

000 **Adapted by Christian theologians:** Salisbury, Joyce E. *The Beast Within: Animals in the Middle Ages*, 2nd ed. (New York: Routledge, 2012).

000 **"eat without pleasure . . . know nothing.":** Jolley, Nicholas. "Malebranche on the Soul." In *The Cambridge Companion to Malebranche*, Steven Nadler, ed. (Cambridge, UK: Cambridge University Press, 2000), 42.

000 **nature-loving landscape architect:** The landscaper was Lynn Mueller, who retired in 2015.

000 **People have done so:** Sax, Boria. *Crow*, new edition. (London: Reaktion Books, 2017).

000 **one line of research:** Jelbert, Sarah A., Alex H. Taylor, and Russell D. Gray. "Investigating Animal Cognition with the Aesop's Fable Paradigm: Current Understanding and Future Directions." *Communicative & Integrative Biology* 8, no. 4 (July 4, 2015): e1035846. https://doi.org/10.1080/19420889.2015.1035846.

000 **adapt their behaviors:** Chow, Pizza Ka Yee, Lisa A. Leaver, Ming Wang, and Stephen E. G. Lea. "Touch Screen Assays of Behavioural Flexibility and Error Characteristics in Eastern Grey Squirrels (*Sciurus carolinensis*)." *Animal Cognition* 20, no. 3 (May 2017): 459–71. https://doi.org/10.1007/s10071-017-1072-z.

000 **Newly hatched chicks . . . rotating cylinder:** Vallortigara, Giorgio, Lucia Regolin, and Fabio Marconato. "Visually Inexperienced Chicks Exhibit Spontaneous Preference for Biological Motion Patterns." *PLoS Biology* 3, no. 7 (June 7, 2005): e208. https://doi.org/10.1371/journal.pbio.0030208.

000 **distinguish between . . . are pushed:** Mascalzoni, Elena, Lucia Regolin, and Giorgio Vallortigara. "Innate Sensitivity for Self-Propelled Causal Agency in Newly Hatched Chicks." *Proceedings of the National Academy of Sciences* 107, no. 9 (March 2, 2010): 4483–85. https://doi.org/10.1073/pnas.0908792107.

000 **Many species possess this ability:** Gonçalves, André, and Dora Biro. "Comparative Thanatology, an Integrative Approach: Exploring Sensory/Cognitive Aspects of Death Recognition in Vertebrates and Invertebrates." *Philosophical Transactions of the Royal Society B: Biological Sciences* 373, no. 1754 (September 5, 2018): 20170263. https://doi.org/10.1098/rstb.2017.0263.

000 **Other research:** Martinho, Antone, and Alex Kacelnik. "Ducklings Imprint on the Relational Concept of 'Same or Different.'" *Science* 353, no. 6296 (July 15, 2016): 286–88. https://doi.org/10.1126/science.aaf4247.

000 **"brutes . . . abstract not.":** Wasserman, Edward A. "Thinking Abstractly like a Duck(Ling)." *Science* 353, no. 6296 (July 15, 2016): 222–23. https://doi.org/10.1126/science.aag3088.

000 **Seven-month-old human infants:** Wasserman, Edward A. "Thinking Abstractly like a Duck(Ling)."

000 **previously completed . . . feed with extra urgency:** Hou, Lily, and Kenneth C. Welch. "Premigratory Ruby-Throated Hummingbirds, *Archilochus colubris*, Exhibit Multiple Strategies for Fuelling Migration." *Animal Behaviour* 121 (November 2016): 87–99. https://doi.org/10.1016/j.anbehav.2016.08.019.

000 **challenged that view:** Breen, Alexis J., Lauren M. Guillette, and Susan D. Healy. "What Can Nest-Building Birds Teach Us?" *Comparative Cognition & Behavior Reviews* 11 (2016): 83–102. https://doi.org/10.3819/ccbr.2016.110005.

000 **learn from experience:** Bailey, Ida E., Kate V. Morgan, Marion Bertin, Simone L. Meddle, and Susan D. Healy. "Physical Cognition: Birds Learn the Structural Effi-

cacy of Nest Material." *Proceedings of the Royal Society B: Biological Sciences* 281, no. 1784 (June 7, 2014): 20133225. https://doi.org/10.1098/rspb.2013.3225.

000 **that personality differences affect metabolism:** Careau, V., D. Thomas, M. M. Humphries, and D. Réale. "Energy Metabolism and Animal Personality." *Oikos* 117, no. 5 (May 2008): 641–53. https://doi.org/10.1111/j.0030-1299.2008.16513.x.

000 **Studies of personality:** Guillette, Lauren M., Marc Naguib, and Andrea S. Griffin. "Individual Differences in Cognition and Personality." *Behavioural Processes* 134 (January 2017): 1–3. https://doi.org/10.1016/j.beproc.2016.12.001.

000 **One recent review:** Delgado, Mikel M., and Frank J. Sulloway. "Attributes of Conscientiousness throughout the Animal Kingdom: An Empirical and Evolutionary Overview." *Psychological Bulletin* 143, no. 8 (August 2017): 823–67. https://doi.org/10.1037/bul0000107.

000 **gave their study a bad reputation:** Wynne, Clive D. L. "What Are Animals? Why Anthropomorphism Is Still Not a Scientific Approach to Behavior." *Comparative Cognition & Behavior Reviews* 2 (2006). https://doi.org/10.3819/ccbr.2008.20008.

000 **"on the intelligence of frogs and toads":** Romanes, George John. *Animal Intelligence.* (New York: D. Appleton, 1892).

000 **extrapolate from past experiences:** Crane, Adam L., and Maud C. O. Ferrari. "Evidence for Risk Extrapolation in Decision Making by Tadpoles." *Scientific Reports* 7, no. 1 (February 23, 2017): 43255. https://doi.org/10.1038/srep43255.

000 **respond . . . less warily:** Crane, Adam L., and Maud C. O. Ferrari. "Evidence for Risk Extrapolation in Decision Making by Tadpoles." *Scientific Reports* 7, no. 1 (February 23, 2017): 43255. https://doi.org/10.1038/srep43255.

000 **learn to discern:** Lucon-Xiccato, Tyrone, Maud C.O. Ferrari, Douglas P. Chivers, and Angelo Bisazza. "Odour Recognition Learning of Multiple Predators by Amphibian Larvae." *Animal Behaviour* 140 (June 2018): 199–205. https://doi.org/10.1016/j.anbehav.2018.04.022.

000 **Italian treefrogs can count:** Lucon-Xiccato, Tyrone, Elia Gatto, and Angelo Bisazza. "Quantity Discrimination by Treefrogs." *Animal Behaviour* 139 (May 2018): 61–69. https://doi.org/10.1016/j.anbehav.2018.03.005.

000 **as high as eight:** Rose, Gary J. "The Numerical Abilities of Anurans and Their Neural Correlates: Insights from Neuroethological Studies of Acoustic Communication." *Philosophical Transactions of the Royal Society B: Biological Sciences* 373, no. 1740 (February 19, 2018): 20160512. https://doi.org/10.1098/rstb.2016.0512.

000 **personality of American bullfrogs:** Carlson, Bradley E., and Tracy Langkilde. "Personality Traits Are Expressed in Bullfrog Tadpoles during Open-Field Trials." *Journal of Herpetology* 47, no. 2 (June 2013): 378–83. https://doi.org/10.1670/12-061.

000 **migrants from southern China:** Song, Ying, Zhenjiang Lan, and Michael H. Kohn. "Mitochondrial DNA Phylogeography of the Norway Rat." *PLoS ONE* 9, no. 2 (February 28, 2014): e88425. https://doi.org/10.1371/journal.pone.0088425.

000 **rats . . . mental time travel:** Corballis, Michael C. "Mental Time Travel: A Case for Evolutionary Continuity." *Trends in Cognitive Sciences* 17, no. 1 (January 2013): 5–6. https://doi.org/10.1016/j.tics.2012.10.009.

000 **the autobiographical sense of self:** Morin, Alain. "Self-Awareness Part 1: Definition, Measures, Effects, Functions, and Antecedents: Self-Awareness." *Social and*

Personality Psychology Compass 5, no. 10 (October 2011): 807–23. https://doi.org/10 .1111/j.1751-9004.2011.00387.x.

000 **Only a few animals . . . :** De Waal, Frans B. M. "Fish, Mirrors, and a Gradualist Perspective on Self-Awareness." *PLOS Biology* 17, no. 2 (February 7, 2019): e3000112. https://doi.org/10.1371/journal.pbio.3000112.

000 **too much emphasis:** Bekoff, Marc. "Awareness: Animal Reflections." *Nature* 419, no. 6904 (September 2002): 255. https://doi.org/10.1038/419255a.

000 **tend not to pass the test:** Broesch, Tanya, Tara Callaghan, Joseph Henrich, Christine Murphy, and Philippe Rochat. "Cultural Variations in Children's Mirror Self-Recognition." *Journal of Cross-Cultural Psychology* 42, no. 6 (August 2011): 1018–29. https://doi.org/10.1177/0022022110381114.

000 **One suggestive experiment found that dogs:** Horowitz, Alexandra. "Smelling Themselves: Dogs Investigate Their Own Odours Longer When Modified in an 'Olfactory Mirror' Test." *Behavioural Processes* 143 (October 2017): 17–24. https://doi .org/10.1016/j.beproc.2017.08.001.

000 **a cousin to this experiment:** Burghardt, Gordon M., Adam M. Partin, Harry E. Pepper, Jordan M. Steele, Samuel M. Liske, Allyson E. Stokes, Ariel N. Lathan, Cary M. Springer, and Matthew S. Jenkins. "Chemically Mediated Self-Recognition in Sibling Juvenile Common Gartersnakes (*Thamnophis sirtalis*) Reared on Same or Different Diets: Evidence for a Chemical Mirror?" *Behaviour* 158, no. 12–13 (September 22, 2021): 1169–91. https://doi.org/10.1163/1568539X-bja10131.

000 **Frans de Waal wrote:** De Waal, Frans B. M. "Fish, Mirrors, and a Gradualist Perspective on Self-Awareness." *PLOS Biology* 17, no. 2 (February 7, 2019): e3000112. https://doi.org/10.1371/journal.pbio.3000112.

000 **Rats are also metacognitive:** Templer, Victoria L., Keith A. Lee, and Aidan J. Preston. "Rats Know When They Remember: Transfer of Metacognitive Responding across Odor-Based Delayed Match-to-Sample Tests." *Animal Cognition* 20, no. 5 (September 2017): 891–906. https://doi.org/10.1007/s10071-017-1109-3.

000 **Episodic-like memory . . . in zebrafish:** Hamilton, Trevor J., Allison Myggland, Erika Duperreault, Zacnicte May, Joshua Gallup, Russell A. Powell, Melike Schalomon, and Shannon M. Digweed. "Episodic-like Memory in Zebrafish." *Animal Cognition* 19, no. 6 (November 2016): 1071–79. https://doi.org/10.1007/s10071-016-1014-1.

000 **this type of memory in hummingbirds:** González-Gómez, Paulina L., Francisco Bozinovic, and Rodrigo A. Vásquez. "Elements of Episodic-like Memory in Free-Living Hummingbirds, Energetic Consequences." *Animal Behaviour* 81, no. 6 (June 2011): 1257–62. https://doi.org/10.1016/j.anbehav.2011.03.014.

000 **mice:** Fellini, Laetitia, and Fabio Morellini. "Mice Create What–Where–When Hippocampus-Dependent Memories of Unique Experiences." *The Journal of Neuroscience* 33, no. 3 (January 16, 2013): 1038–43. https://doi.org/10.1523/JNEUROSCI .2280-12.2013.

000 **cuttlefish:** Jozet-Alves, Christelle, Marion Bertin, and Nicola S. Clayton. "Evidence of Episodic-like Memory in Cuttlefish." *Current Biology* 23, no. 23 (December 2013): R1033–35. https://doi.org/10.1016/j.cub.2013.10.021.

000 **They distinguish between:** Godfrey-Smith, Peter. "The evolution of consciousness in phylogenetic context." In *The Routledge Handbook of Philosophy of Animal Minds*, Kristin Andrews and Jacob Beck, eds. (New York: Routledge, 2017), 216–26.

000 **Forms of it have existed ever since:** Lacalli, Thurston. "Amphioxus Neurocir-

cuits, Enhanced Arousal, and the Origin of Vertebrate Consciousness." *Consciousness and Cognition* 62 (July 2018): 127–34. https://doi.org/10.1016/j.concog.2018.03.006.

000 **It's why an earthworm responds differently:** Merker, Bjorn. "The Liabilities of Mobility: A Selection Pressure for the Transition to Consciousness in Animal Evolution." *Consciousness and Cognition* 14, no. 1 (March 2005): 89–114. https://doi.org/10.1016/S1053-8100(03)00002-3.

000 **an experiment in which recently fed rat snakes:** Khvatov, Sokolov, and Kharitonov. "Snakes *Elaphe Radiata* May Acquire Awareness of Their Body Limits When Trying to Hide in a Shelter." *Behavioral Sciences* 9, no. 7 (June 26, 2019): 67. https://doi.org/10.3390/bs9070067.

000 **the movements of jewel-eyed jumping spiders:** Jürgens, Uta M. "Universal Modes of Awareness? A 'Pre-Reflective' Premise." *Animal Sentience* 1, no. 10 (August 11, 2016). https://doi.org/10.51291/2377-7478.1129.

000 **writes philosopher Mark Rowlands:** Rowlands, Mark. "Are Animals Persons?" *Animal Sentience* 1, no. 10 (July 12, 2016). https://doi.org/10.51291/2377-7478.1110.

000 **they can plan . . . a half-hour ahead:** Feeney, Miranda C., William A. Roberts, and David F. Sherry. "Black-Capped Chickadees (*Poecile atricapillus*) Anticipate Future Outcomes of Foraging Choices." *Journal of Experimental Psychology: Animal Behavior Processes* 37, no. 1 (2011): 30–40. https://doi.org/10.1037/a0019908.

2. LANDSCAPES OF THE HEART

000 **the largest . . . in the county:** "The National Institutes of Health's Green Features." National Institutes of Health Office of Management, 2015.

000 **a controversial statement:** Bekoff, Marc. *The Emotional Lives of Animals: A Leading Scientist Explores Animal Joy, Sorrow, and Empathy—and Why They Matter.* (Novato, CA: New World Library, 2007).

000 **a sharp contrast between intelligence and emotion:** Berridge, Kent C., and Morten L. Kringelbach. "Affective Neuroscience of Pleasure: Reward in Humans and Animals." *Psychopharmacology* 199, no. 3 (August 2008): 457–80. https://doi.org/10.1007/s00213-008-1099-6.

000 **Plato located rationality and passion:** "Plato's Three Parts of the Soul." Accessed June 28, 2023. https://philosophycourse.info/platosite/3schart.html.

000 **experiencing emotions . . . required language:** Lindquist, Kristen A., Ajay B. Satpute, and Maria Gendron. "Does Language Do More than Communicate Emotion?" *Current Directions in Psychological Science* 24, no. 2 (April 2015): 99–108. https://doi.org/10.1177/0963721414553440.

000 **we simply could not determine:** Harnad, Stevan. "Animal Sentience: The Other-Minds Problem." *Animal Sentience* 1, no. 1 (January 1, 2016). https://doi.org/10.51291/2377-7478.1065.

000 **Darwin wrote . . . about a Sulawesi macaque:** Waal, Frans B. M. "Darwin's Legacy and the Study of Primate Visual Communication." *Annals of the New York Academy of Sciences* 1000, no. 1 (January 24, 2006): 7–31. https://doi.org/10.1196/annals.1280.003.

000 **cognition and emotion . . . as intertwined:** Pessoa, Luiz, Loreta Medina, Patrick R. Hof, and Ester Desfilis. "Neural Architecture of the Vertebrate Brain: Im-

plications for the Interaction between Emotion and Cognition." *Neuroscience & Biobehavioral Reviews* 107 (December 2019): 296–312. https://doi.org/10.1016/j .neubiorev.2019.09.021.

000 **evolved with the earliest vertebrates:** O'Connell, Lauren A., and Hans A. Hofmann. "The Vertebrate Mesolimbic Reward System and Social Behavior Network: A Comparative Synthesis." *The Journal of Comparative Neurology* 519, no. 18 (December 15, 2011): 3599–3639. https://doi.org/10.1002/cne.22735.

000 **voles are used to study the neurobiology of love:** Manoli, Dev, Steven Phelps, and Zoe Donaldson. "Monogamous Prairie Voles Reveal the Neurobiology of Love." *Scientific American*, February 1, 2023.

000 **salmon become frustrated:** Vindas, Marco A., Ole Folkedal, Tore S. Kristiansen, Lars H. Stien, Bjarne O. Braastad, Ian Mayer, and Øyvind Øverli. "Omission of Expected Reward Agitates Atlantic Salmon (*Salmo salar*)." *Animal Cognition* 15, no. 5 (September 2012): 903–11. https://doi.org/10.1007/s10071-012-0517-7.

000 **crawfish experience anxiety:** Perry, Clint J., and Luigi Baciadonna. "Studying Emotion in Invertebrates: What Has Been Done, What Can Be Measured and What They Can Provide." *Journal of Experimental Biology* 220, no. 21 (November 1, 2017): 3856–68. https://doi.org/10.1242/jeb.151308.

000 **a suggestive study on rats:** Cabanac, M., and K. G. Johnson. "Analysis of a Conflict between Palatability and Cold Exposure in Rats." *Physiology & Behavior* 31, no. 2 (August 1983): 249–53. https://doi.org/10.1016/0031-9384(83)90128-2.

000 **It helps them . . . learn about the world:** Bateson, Patrick. "Playfulness and Creativity." *Current Biology* 25, no. 1 (January 2015): R12–16. https://doi.org/10.1016/j .cub.2014.09.009.

000 **rules of social etiquette:** Ward, Camille, Erika B. Bauer, and Barbara B. Smuts. "Partner Preferences and Asymmetries in Social Play among Domestic Dog, *Canis lupus familiaris*, Littermates." *Animal Behaviour* 76, no. 4 (October 2008): 1187–99. https://doi.org/10.1016/j.anbehav.2008.06.004.

000 **flush of dopamine should reward:** Extrapolation from the presence of dopamine in squirrel brains and its observed role in the play of other rodents, e.g., Vanderschuren, Louk J. M. J., E. J. Marijke Achterberg, and Viviana Trezza. "The Neurobiology of Social Play and Its Rewarding Value in Rats." *Neuroscience & Biobehavioral Reviews* 70 (November 2016): 86–105. https://doi.org/10.1016/j.neubiorev.2016.07.025.

000 **which scientists formally define:** Burghardt, Gordon M. "Play in Fishes, Frogs and Reptiles." *Current Biology* 25, no. 1 (January 2015): R9–10. https://doi.org/10 .1016/j.cub.2014.10.027.

000 **Vietnamese mossy frog tadpoles catch rides:** Burghardt, Gordon M. "Play in Fishes, Frogs and Reptiles." *Current Biology* 25, no. 1 (January 2015): R9–10. https://doi.org/10.1016/j.cub.2014.10.027.

000 **elephantfish balance twigs:** Stefoff, Rebecca. *How Animals Play.* (New York: Cavendish Square), 2013.

000 **Komodo dragons lumbering after a ball:** Burghardt, Gordon M. "Play in Fishes, Frogs and Reptiles." *Current Biology* 25, no. 1 (January 2015): R9–10. https://doi .org/10.1016/j.cub.2014.10.027

000 **Melissa Bateson conducted a series of experiments:** Brilot, Ben O., and Melissa Bateson. "Water Bathing Alters Threat Perception in Starlings." *Biology Letters* 8, no. 3 (June 23, 2012): 379–81. https://doi.org/10.1098/rsbl.2011.1200.

000 **the psychological effects of close confinement on pigs:** Scollo, Annalisa, Flaviana Gottardo, Barbara Contiero, and Sandra A. Edwards. "Does Stocking Density Modify Affective State in Pigs as Assessed by Cognitive Bias, Behavioural and Physiological Parameters?" *Applied Animal Behaviour Science* 153 (April 2014): 26–35. https://doi.org/10.1016/j.applanim.2014.01.006.

000 **whether sniffing makes dogs happy:** Duranton, C., and A. Horowitz. "Let Me Sniff! Nosework Induces Positive Judgment Bias in Pet Dogs." *Applied Animal Behaviour Science* 211 (February 2019): 61–66. https://doi.org/10.1016/j.applanim.2018.12.009.

000 **ravens become upset:** Adriaense, Jessie E. C., Jordan S. Martin, Martina Schiestl, Claus Lamm, and Thomas Bugnyar. "Negative Emotional Contagion and Cognitive Bias in Common Ravens (*Corvus corax*)." *Proceedings of the National Academy of Sciences* 116, no. 23 (June 4, 2019): 11547–52. https://doi.org/10.1073/pnas.1817066116.

000 **refined these tests for bees:** Bateson, Melissa, Suzanne Desire, Sarah E. Gartside, and Geraldine A. Wright. "Agitated Honeybees Exhibit Pessimistic Cognitive Biases." *Current Biology* 21, no. 12 (June 2011): 1070–73. https://doi.org/10.1016/j.cub.2011.05.017.

000 **similar work with bumblebees:** Solvi, Cwyn, Luigi Baciadonna, and Lars Chittka. "Unexpected Rewards Induce Dopamine-Dependent Positive Emotion-like State Changes in Bumblebees." *Science* 353, no. 6307 (September 30, 2016): 1529–31. https://doi.org/10.1126/science.aaf4454.

000 **the bees no longer act emotional:** Perry, Clint J., and Luigi Baciadonna. "Studying Emotion in Invertebrates: What Has Been Done, What Can Be Measured and What They Can Provide." *Journal of Experimental Biology* 220, no. 21 (November 1, 2017): 3856–68. https://doi.org/10.1242/jeb.151308.

000 **brain structures and chemistries:** Perry, "Studying Emotion in Invertebrates: What Has Been Done, What Can Be Measured and What They Can Provide."

000 **fruit flies enjoy ejaculation:** Zer-Krispil, Shir, Hila Zak, Lisha Shao, Shir Ben-Shaanan, Lea Tordjman, Assa Bentzur, Anat Shmueli, and Galit Shohat-Ophir. "Ejaculation Induced by the Activation of Crz Neurons Is Rewarding to *Drosophila* Males." *Current Biology* 28, no. 9 (May 2018): 1445–52, e3. https://doi.org/10.1016/j.cub.2018.03.039.

000 **when subject to repeated, uncontrollable stresses:** Yang, Zhenghong, Franco Bertolucci, Reinhard Wolf, and Martin Heisenberg. "Flies Cope with Uncontrollable Stress by Learned Helplessness." *Current Biology* 23, no. 9 (May 2013): 799–803.

000 **an old English derivation of *folk*:** Talbot, William Henry Fox. *English Etymologies*. (London: John Murray, 1847).

000 **pain, that union of sensation and emotion:** Gilam, Gadi, James J. Gross, Tor D. Wager, Francis J. Keefe, and Sean C. Mackey. "What Is the Relationship between Pain and Emotion? Bridging Constructs and Communities." *Neuron* 107, no. 1 (July 2020): 17–21. https://doi.org/10.1016/j.neuron.2020.05.024.

000 **Until fairly recently, it was argued:** National Research Council (US) Committee on Recognition and Alleviation of Pain in Laboratory Animals. *Pain in Research Animals: General Principles and Considerations.* (Washington, DC: National Academies Press), 2009. https://www.ncbi.nlm.nih.gov/books/NBK32655/.

000 **studies show that . . . fish behave as we would:** Sneddon, Lynne U. "Evolution of

Nociception and Pain: Evidence from Fish Models." *Philosophical Transactions of the Royal Society B: Biological Sciences* 374, no. 1785 (November 11, 2019): 20190290. https://doi.org/10.1098/rstb.2019.0290.

000 **The pain-sensing neurons of trout:** Sneddon, "Evolution of Nociception and Pain: Evidence from Fish Models."

000 **amphibians can perceive faint electric fields:** Crampton, William G. R. "Electroreception, Electrogenesis and Electric Signal Evolution." *Journal of Fish Biology* 95, no. 1 (July 2019): 92–134. https://doi.org/10.1111/jfb.13922.

000 **is underappreciated:** Valentine, Katie. "Female Birdsong Is Finally Getting the Attention It Deserves," *Audubon,* August 15, 2017.

000 **wrote the nineteenth-century poet Sidney Lanier:** Cameron, Gabe. "The Establishment and Development of the Mockingbird as the Nightingale's 'American Rival.'" *Electronic Theses and Dissertations,* May 1, 2017. https://dc.etsu.edu/etd/3224.

000 **also make their own, original melodies:** "A Mockingbird Acquires His Song Repertory." *The Auk* 61, no. 2 (April 1944): 211–19. https://doi.org/10.2307/4079364.

000 **In New Orleans, disagreement exists:** Abrams, Eve. "Who Sang It First? Mockingbirds and Musicians Cover Each Other in New Orleans." *NPR,* October 23, 2014.

000 **beating their bounds, a millennia-old tradition:** Winterhalter, Elizabeth. "'Beating the Bounds.'" *JSTOR Daily,* May 7, 2020. https://daily.jstor.org/beating-the-bounds/.

000 **their unions last for years . . . :** Logan, Cheryl A. "Mate Re-assessment in an Already-Mated Female Northern Mockingbird." *Chat* 61 (1997): 108–12.

000 **Research on zebra finches . . . :** Riters, Lauren V., Brandon J. Polzin, Alyse N. Maksimoski, Sharon A. Stevenson, and Sarah J. Alger. "Birdsong and the Neural Regulation of Positive Emotion." *Frontiers in Psychology* 13 (June 22, 2022): 903857. https://doi.org/10.3389/fpsyg.2022.903857.

000 **a study of white-throated sparrows:** Earp, Sarah E., and Donna L. Maney. "Birdsong: Is It Music to Their Ears?" *Frontiers in Evolutionary Neuroscience* 4 (2012). https://doi.org/10.3389/fnevo.2012.00014.

000 **basic capacities . . . he believed to be widespread among animals:** Darwin, Charles. *The Expression of the Emotions in Man and Animals.* (London: John Murray, 1872).

000 **a proposition tested several years ago:** Filippi, Piera, Jenna V. Congdon, John Hoang, Daniel L. Bowling, Stephan A. Reber, Andrius Pašukonis, Marisa Hoeschele, et al. "Humans Recognize Emotional Arousal in Vocalizations across All Classes of Terrestrial Vertebrates: Evidence for Acoustic Universals." *Proceedings of the Royal Society B: Biological Sciences* 284, no. 1859 (July 26, 2017): 20170990. https://doi.org/10.1098/rspb.2017.0990.

000 **Other researchers analyzed:** Lingle, Susan, Megan T. Wyman, Radim Kotrba, Lisa J. Teichroeb, and Cora A. Romanow. "What Makes a Cry a Cry? A Review of Infant Distress Vocalizations." *Current Zoology* 58, no. 5 (October 1, 2012): 698–726. https://doi.org/10.1093/czoolo/58.5.698.

000 **recordings of distressed baby mammals . . . are played:** Lingle, Susan, and Tobias Riede. "Deer Mothers Are Sensitive to Infant Distress Vocalizations of Diverse Mammalian Species." *The American Naturalist* 184, no. 4 (October 2014): 510–22. https://doi.org/10.1086/677677.

000 **Findings like these:** Congdon, Jenna V., Allison H. Hahn, Piera Filippi, Kimberley A. Campbell, John Hoang, Erin N. Scully, Daniel L. Bowling, Stephan A. Reber, and Christopher B. Sturdy. "Hear Them Roar: A Comparison of Black-Capped Chickadee (*Poecile atricapillus*) and Human (*Homo sapiens*) Perception of Arousal in Vocalizations across All Classes of Terrestrial Vertebrates." *Journal of Comparative Psychology* 133, no. 4 (November 2019): 520–41. https://doi.org/10.1037/com0000187.

000 **the neurobiological networks . . . are intertwined:** O'Connell, Lauren A., and Hans A. Hofmann. "The Vertebrate Mesolimbic Reward System and Social Behavior Network: A Comparative Synthesis." *The Journal of Comparative Neurology* 519, no. 18 (December 15, 2011): 3599–3639. https://doi.org/10.1002/cne.22735.

000 **how deeply rats feel for one another:** Meyza, K. Z., I. Ben-Ami Bartal, M. H. Monfils, J. B. Panksepp, and E. Knapska. "The Roots of Empathy: Through the Lens of Rodent Models." *Neuroscience & Biobehavioral Reviews* 76 (May 2017): 216–34. https://doi.org/10.1016/j.neubiorev.2016.10.028.

000 **evolved to foster sociality:** Decety, Jean, Inbal Ben-Ami Bartal, Florina Uzefovsky, and Ariel Knafo-Noam. "Empathy as a Driver of Prosocial Behaviour: Highly Conserved Neurobehavioural Mechanisms across Species." *Philosophical Transactions of the Royal Society B: Biological Sciences* 371, no. 1686 (January 19, 2016): 20150077. https://doi.org/10.1098/rstb.2015.0077.

000 **associated with parental care:** Cunningham, Christopher B., Majors J. Badgett, Richard B. Meagher, Ron Orlando, and Allen J. Moore. "Ethological Principles Predict the Neuropeptides Co-opted to Influence Parenting." *Nature Communications* 8, no. 1 (February 1, 2017): 14225. https://doi.org/10.1038/ncomms14225.

000 **a hormone central to . . . reconciliation and affection:** Burkett, J. P., E. Andari, Z. V. Johnson, D. C. Curry, F. B. M. De Waal, and L. J. Young. "Oxytocin-Dependent Consolation Behavior in Rodents." *Science* 351, no. 6271 (January 22, 2016): 375–78. https://doi.org/10.1126/science.aac4785.

000 **plays a part in . . . every vertebrate class:** Decety, "Empathy as a Driver of Prosocial Behaviour: Highly Conserved Neurobehavioural Mechanisms across Species."

000 **a review of this behavior:** Kenny, Elspeth, Tim R. Birkhead, and Jonathan P. Green. "Allopreening in Birds Is Associated with Parental Cooperation over Offspring Care and Stable Pair Bonds across Years." *Behavioral Ecology* 28, no. 4 (August 1, 2017): 1142–48. https://doi.org/10.1093/beheco/arx078.

000 **Such rituals:** Roth, Tom S., Iliana Samara, Jingzhi Tan, Eliska Prochazkova, and Mariska E. Kret. "A Comparative Framework of Inter-individual Coordination and Pair-Bonding." *Current Opinion in Behavioral Sciences* 39 (June 2021): 98–105. https://doi.org/10.1016/j.cobeha.2021.03.005.

000 **less stressed by fights:** Wascher, Claudia A. F., Brigitte M. Weiß, Walter Arnold, and Kurt Kotrschal. "Physiological Implications of Pair-Bond Status in Greylag Geese." *Biology Letters* 8, no. 3 (June 23, 2012): 347–50. https://doi.org/10.1098/rsbl.2011.0917.

000 **become upset when separated:** Ludwig, Sonja C., Katharina Kapetanopoulos, Kurt Kotrschal, and Claudia A.F. Wascher. "Effects of Mate Separation in Female and Social Isolation in Male Free-Living Greylag Geese on Behavioural and Phys-

iological Measures." *Behavioural Processes* 138 (May 2017): 134–41. https://doi.org/10.1016/j.beproc.2017.03.002.

000 **"Geese possess a veritably human capacity for grief":** Lorenz, Konrad. *The Year of the Greylag Goose*. (New York: Harcourt Brace Jovanovich, 1979).

000 **A Canada goose who returned daily:** Lorenz, Konrad. *The Year of the Greylag Goose*.

000 **The scientific explanation . . . ensured closer cooperation:** Black, J. M. "Fitness Consequences of Long-Term Pair Bonds in Barnacle Geese: Monogamy in the Extreme." *Behavioral Ecology* 12, no. 5 (September 1, 2001): 640–45. https://doi.org/10.1093/beheco/12.5.640.

000 **relatively uncommon in the animal kingdom:** Ryan, Michael J., Rachel A. Page, Kimberly L. Hunter, and Ryan C. Taylor. " 'Crazy Love': Nonlinearity and Irrationality in Mate Choice." *Animal Behaviour* 147 (January 2019): 189–98. https://doi.org/10.1016/j.anbehav.2018.04.004.

000 **feel good to gaze upon:** Skov, Martin, and Marcos Nadal. "The Nature of Beauty: Behavior, Cognition, and Neurobiology." *Annals of the New York Academy of Sciences* 1488, no. 1 (March 2021): 44–55. https://doi.org/10.1111/nyas.14524.

3. NO ANIMAL IS AN ISLAND

000 **biologists . . . did just that:** Templeton, Christopher N., Veronica A. Reed, S. Elizabeth Campbell, and Michael D. Beecher. "Spatial Movements and Social Networks in Juvenile Male Song Sparrows." *Behavioral Ecology* 23, no. 1 (January 1, 2012): 141–52. https://doi.org/10.1093/beheco/arr167.

000 **the experience of anthropologist Barbara Smuts:** Interview with Barbara Smuts.

000 **Female barnacle geese . . . prefer the company:** Kurvers, Ralf H. J. M., Lea Prox, Damien R. Farine, Coretta Jongeling, and Lysanne Snijders. "Season-Specific Carryover of Early Life Associations in a Monogamous Bird Species." *Animal Behaviour* 164 (June 2020): 25–37. https://doi.org/10.1016/j.anbehav.2020.03.016.

000 **starlings . . . tend to be polygynous:** Henry, Laurence, Cécile Bourguet, Marion Coulon, Christine Aubry, and Martine Hausberger. "Sharing Mates and Nest Boxes Is Associated with Female 'Friendship' in European Starlings, *Sturnus vulgaris*." *Journal of Comparative Psychology* 127, no. 1 (2013): 1–13. https://doi.org/10.1037/a0029975.

000 **one spectacular account of migration involves European bee-eaters:** Dhanjal-Adams, Kiran L., Silke Bauer, Tamara Emmenegger, Steffen Hahn, Simeon Lisovski, and Felix Liechti. "Spatiotemporal Group Dynamics in a Long-Distance Migratory Bird." *Current Biology* 28, no. 17 (September 2018): 2824–2830.e3. https://doi.org/10.1016/j.cub.2018.06.054.

000 **certain garter snakes have an affinity:** Skinner, Morgan, and Noam Miller. "Aggregation and Social Interaction in Garter Snakes (*Thamnophis sirtalis sirtalis*)." *Behavioral Ecology and Sociobiology* 74, no. 5 (May 2020): 51. https://doi.org/10.1007/s00265-020-2827-0.

000 **Females are less aggressive:** Vickruck, J. L., and M. H. Richards. "Nestmate Discrimination Based on Familiarity but Not Relatedness in Eastern Carpenter

Bees." *Behavioural Processes* 145 (December 2017): 73–80. https://doi.org/10.1016/j.beproc.2017.10.005.

000 **sunfish . . . able to recognize one another:** Dugatkin, Lee Alan, and David Sloan Wilson. "The Prerequisites for Strategic Behaviour in Bluegill Sunfish, *Lepomis macrochirus.*" *Animal Behaviour* 44 (August 1992): 223–30. https://doi.org/10.1016/0003-3472(92)90028-8.

000 **one-eighth of the state's population:** Bennage, Sebastian. "Maine Sets a New Record This Hunting Season." *News Center Maine*, December 7, 2022.

000 **Still, they move between seasonal territories:** Nelson, Michael E. "Home Range Location of White-Tailed Deer." United States: Department of Agriculture, Forest Service, North Central Forest Experiment Station, 1979.

000 **research now suggests:** Jesmer, Brett R., Jerod A. Merkle, Jacob R. Goheen, Ellen O. Aikens, Jeffrey L. Beck, Alyson B. Courtemanch, Mark A. Hurley, et al. "Is Ungulate Migration Culturally Transmitted? Evidence of Social Learning from Translocated Animals." *Science* 361, no. 6406 (September 7, 2018): 1023–25. https://doi.org/10.1126/science.aat0985.

000 **Knowledge . . . is part of their culture:** Nelson, Michael E, and L David Mech. "Twenty-Year Home-Range Dynamics of a White-Tailed Deer Matriline." *Canadian Journal of Zoology* 77, no. 7 (October 20, 1999): 1128–35. https://doi.org/10.1139/z99-085.

000 **an Australian community of bottlenose dolphins:** Wild, Sonja, Simon J. Allen, Michael Krützen, Stephanie L. King, Livia Gerber, and William J. E. Hoppitt. "Multi-Network-Based Diffusion Analysis Reveals Vertical Cultural Transmission of Sponge Tool Use within Dolphin Matrilines." *Biology Letters* 15, no. 7 (July 2019): 20190227. https://doi.org/10.1098/rsbl.2019.0227.

000 **greeting rituals of white-faced capuchin monkeys:** Perry, Susan. "Social Traditions and Social Learning in Capuchin Monkeys (*Cebus*)." *Philosophical Transactions of the Royal Society B: Biological Sciences* 366, no. 1567 (April 12, 2011): 988–96. https://doi.org/10.1098/rstb.2010.0317.

000 **carrion crows in Japan:** Nihei, Yoshiaki. "Variations of Behaviour of Carrion Crows *Corvus corone* Using Automobiles as Nutcrackers." *Japanese Journal of Ornithology* 44, no. 1 (1995): 21–35. https://doi.org/10.3838/jjo.44.21.

000 **bumblebees . . . learning to solve puzzles:** Bridges, Alice D., HaDi MaBouDi, Olga Procenko, Charlotte Lockwood, Yaseen Mohammed, Amelia Kowalewska, José Eric Romero González, Joseph L. Woodgate, and Lars Chittka. "Bumblebees Acquire Alternative Puzzle-Box Solutions via Social Learning." *PLOS Biology* 21, no. 3 (March 7, 2023): e3002019. https://doi.org/10.1371/journal.pbio.3002019.

000 **identify blossoms:** Baracchi, David, Vera Vasas, Soha Jamshed Iqbal, and Sylvain Alem. "Foraging Bumblebees Use Social Cues More When the Task Is Difficult." *Behavioral Ecology* 29, no. 1 (January 13, 2018): 186–92. https://doi.org/10.1093/beheco/arx143.

000 **demonstrated in whooping cranes:** Mueller, Thomas, Robert B. O'Hara, Sarah J. Converse, Richard P. Urbanek, and William F. Fagan. "Social Learning of Migratory Performance." *Science* 341, no. 6149 (August 30, 2013): 999–1002. https://doi.org/10.1126/science.1237139.

000 **one recent review of animal migration:** Aikens, Ellen O., Iris D. Bontekoe, Lara Blumenstiel, Anna Schlicksupp, and Andrea Flack. "Viewing Animal Migration

through a Social Lens." *Trends in Ecology & Evolution* 37, no. 11 (November 2022): 985–96. https://doi.org/10.1016/j.tree.2022.06.008.

000 **the seminal description of animal culture:** Fisher, James and Hinde, R. A. "The Opening of Milk Bottles by Birds." *British Birds* 42 (1949): 347–57.

000 **suggested that house sparrows:** Liker, András, and Veronika Bókony. "Larger Groups Are More Successful in Innovative Problem Solving in House Sparrows." *Proceedings of the National Academy of Sciences* 106, no. 19 (May 12, 2009): 7893–98. https://doi.org/10.1073/pnas.0900042106.

000 **and pigeons:** Skandrani, Zina, Dalila Bovet, Julien Gasparini, Natale Emilio Baldaccini, and Anne-Caroline Prévot. "Sociality Enhances Birds' Capacity to Deal with Anthropogenic Ecosystems." *Urban Ecosystems* 20, no. 3 (June 2017): 609–15. https://doi.org/10.1007/s11252-016-0618-1.

000 **biologist Peter Marler discovered:** Marler, P., and M. Tamura. "Song 'Dialects' in Three Populations of White-Crowned Sparrows." *The Condor* 64, no. 5 (September 1962): 368–77. https://doi.org/10.2307/1365545.

000 **Dialects have since been described:** Henry, Laurence, Stéphanie Barbu, Alban Lemasson, and Martine Hausberger. "Dialects in Animals: Evidence, Development and Potential Functions." *Animal Behavior and Cognition* 2, no. 2 (May 1, 2015): 132–55. https://doi.org/10.12966/abc.05.03.2015.

000 **vervet monkeys had specific alarms:** Seyfarth, Robert M., Dorothy L. Cheney, and Peter Marler. "Monkey Responses to Three Different Alarm Calls: Evidence of Predator Classification and Semantic Communication." *Science* 210, no. 4471 (November 14, 1980): 801–3. https://doi.org/10.1126/science.7433999.

000 **Marler's obituary:** Vitello, Paul. "Peter Marler, Graphic Decoder of Birdsong, Dies at 86." *New York Times*, July 28, 2014.

000 **language . . . unique to humans:** Nimmo, Richie. "From over the horizon: Animal alterity and liminal intimacy beyond the anthropomorphic embrace." *Otherness: Essays and Studies* 5, no. 2 (2016).

000 **no animal can share:** Lindquist, Kristen A., Ajay B. Satpute, and Maria Gendron. "Does Language Do More Than Communicate Emotion?" *Current Directions in Psychological Science* 24, no. 2 (April 2015): 99–108. https://doi.org/10.1177/0963721414553440.

000 **The essence of communication is information exchange:** Schmidt, Kenneth A., Sasha R. X. Dall, and Jan A. Van Gils. "The Ecology of Information: An Overview on the Ecological Significance of Making Informed Decisions." *Oikos* 119, no. 2 (February 2010): 304–16. https://doi.org/10.1111/j.1600-0706.2009.17573.x.

000 **a far more nuanced understanding:** Horowitz, Alexandra. *Being a Dog: Following the Dog Into a World of Smell.* (New York: Scribner's, 2016).

000 **beavers do so:** Aleksiuk, M. "Scent-Mound Communication, Territoriality, and Population Regulation in Beaver (*Castor canadensis Kuhl*)." *Journal of Mammalogy* 49, no. 4 (November 26, 1968): 759–62. https://doi.org/10.2307/1378741.

000 **river otters:** Barocas, Adi, Howard N. Golden, Michael W. Harrington, David B. McDonald, and Merav Ben-David. "Coastal Latrine Sites as Social Information Hubs and Drivers of River Otter Fission–Fusion Dynamics." *Animal Behaviour* 120 (October 2016): 103–14. https://doi.org/10.1016/j.anbehav.2016.07.016.

000 **bears:** Taylor, A. Preston, Maximilian L. Allen, and Micaela S. Gunther. "Black Bear Marking Behaviour at Rub Trees during the Breeding Season in Northern Califor-

nia." *Behaviour* 152, no. 7–8 (2015): 1097–1111. https://doi.org/10.1163/1568539X
-00003270.

000 **house mice:** Hurst, Jane L. "Scent Marking and Social Communication." In *Animal Communication Networks*, P. K. McGregor, ed. (Cambridge, UK: Cambridge University Press, 2005), 219–44. https://doi.org/10.1017/CBO9780511610363.014.

000 **moose:** Bowyer, R. Terry, Victor Van Ballenberghe, and Karen R. Rock. "Scent Marking by Alaskan Moose: Characteristics and Spatial Distribution of Rubbed Trees." *Canadian Journal of Zoology* 72, no. 12 (December 1, 1994): 2186–92. https://doi.org/10.1139/z94-292.

000 **garter snakes . . . lay a chemical trail:** Costanzo, Jon P. "Conspecific Scent Trailing by Garter Snakes (*Thamnophis sirtalis*) during Autumn Further Evidence for Use of Pheromones in Den Location." *Journal of Chemical Ecology* 15, no. 11 (November 1989): 2531–38. https://doi.org/10.1007/BF01014729.

000 **One study of chimpanzees:** Bard, Kim A., Vanessa Maguire-Herring, Masaki Tomonaga, and Tetsuro Matsuzawa. "The Gesture 'Touch': Does Meaning-Making Develop in Chimpanzees' Use of a Very Flexible Gesture?" *Animal Cognition* 22, no. 4 (July 2019): 535–50. https://doi.org/10.1007/s10071-017-1136-0.

000 **A squirrel's semaphore-like tail-waving:** McRae, Thaddeus R., and Steven M. Green. "Joint Tail and Vocal Alarm Signals of Gray Squirrels (*Sciurus carolinensis*)." *Behaviour* 151, no. 10 (2014): 1433–52. https://doi.org/10.1163/1568539X
-00003194.

000 **crayfish waving their antennae:** Bruski, Colleen Ann, and D. W. Dunham. "The Importance of Vision in Agonistic Communication of the Crayfish Orconectes Rusticus. I: An Analysis of Bout Dynamics." *Behaviour* 103, no. 1–3 (1987): 83–107. https://doi.org/10.1163/156853987X00288.

000 **northern snake-necked turtles . . . vocalizations:** Giles, Jacqueline C., Jenny A. Davis, Robert D. McCauley, and Gerald Kuchling. "Voice of the Turtle: The Underwater Acoustic Repertoire of the Long-Necked Freshwater Turtle, *Chelodina oblonga*." *The Journal of the Acoustical Society of America* 126, no. 1 (July 2009): 434–43. https://doi.org/10.1121/1.3148209.

000 **used by South American giant river turtles:** Ferrara, Camila R., Richard C. Vogt, and Renata S. Sousa-Lima. "Turtle Vocalizations as the First Evidence of Posthatching Parental Care in Chelonians." *Journal of Comparative Psychology* 127, no. 1 (2013): 24–32. https://doi.org/10.1037/a0029656.

000 **chickens telling flockmates about food:** Evans, Christopher S., and Linda Evans. "Chicken Food Calls Are Functionally Referential." *Animal Behaviour* 58, no. 2 (August 1999): 307–19. https://doi.org/10.1006/anbe.1999.1143.

000 **research conducted by ethologist Toshitaka Suzuki:** Suzuki, Toshitaka N., David Wheatcroft, and Michael Griesser. "Experimental Evidence for Compositional Syntax in Bird Calls." *Nature Communications* 7, no. 1 (March 8, 2016): 10986. https://doi.org/10.1038/ncomms10986.

000 **What's more, Suzuki showed:** Suzuki, Toshitaka N. "Imagery in Wild Birds: Retrieval of Visual Information from Referential Alarm Calls." *Learning & Behavior* 47, no. 2 (June 2019): 111–14. https://doi.org/10.3758/s13420-019-00374-9.

000 **chickadee call . . . quite complex:** Hahn, Allison H., Jenna V. Congdon, Kimberley A. Campbell, Erin N. Scully, Neil McMillan, and Christopher B. Sturdy. "Mechanisms of Communication and Cognition in Chickadees." *Advances in the*

Study of Behavior, 49:147–97. Elsevier, 2017. https://doi.org/10.1016/bs.asb.2017.02.003.

000 **as common murres do:** Storey, Anne E., Sabina I. Wilhelm, and Carolyn J. Walsh. "Negotiation of Parental Duties in Chick-Rearing Common Murres (*Uria aalge*) in Different Foraging Conditions." *Frontiers in Ecology and Evolution* 7 (January 14, 2020): 506. https://doi.org/10.3389/fevo.2019.00506.

000 **many species . . . respond to the alarm calls:** Carlson, Nora V., Erick Greene, and Christopher N. Templeton. "Nuthatches Vary Their Alarm Calls Based upon the Source of the Eavesdropped Signals." *Nature Communications* 11, no. 1 (January 27, 2020): 526. https://doi.org/10.1038/s41467-020-14414-w.

000 **Groundhogs listen to . . . chipmunks:** Aschemeier, Lisa M., and Christine R. Maher. "Eavesdropping of Woodchucks (*Marmota monax*) and Eastern Chipmunks (*Tamias striatus*) on Heterospecific Alarm Calls." *Journal of Mammalogy* 92, no. 3 (June 9, 2011): 493–99. https://doi.org/10.1644/09-MAMM-A-322.1.

000 **they're exchanging information:** Keim, Brandon. "Decoding Nature's Soundtrack." *Nautilus*, April 7, 2014.

000 **a point raised by . . . Leonie Cornips:** Cornips, Leonie. "The Final Frontier: Non-Human Animals on the Linguistic Research Agenda." *Linguistics in the Netherlands* 36 (November 5, 2019): 13–19. https://doi.org/10.1075/avt.00015.cor.

000 **sociality is the great driver of intelligence:** Emery, Nathan J., Nicola S. Clayton, and Chris D. Frith. "Introduction. Social Intelligence: From Brain to Culture." *Philosophical Transactions of the Royal Society B: Biological Sciences* 362, no. 1480 (April 29, 2007): 485–88. https://doi.org/10.1098/rstb.2006.2022.

000 **great apes do:** Krupenye, Christopher, Fumihiro Kano, Satoshi Hirata, Josep Call, and Michael Tomasello. "Great Apes Anticipate That Other Individuals Will Act According to False Beliefs." *Science* 354, no. 6308 (October 7, 2016): 110–14. https://doi.org/10.1126/science.aaf8110.

000 **evidence for cetaceans:** Tomonaga, Masaki, and Yuka Uwano. "Bottlenose Dolphins' (*Tursiops truncatus*) Theory of Mind as Demonstrated by Responses to Their Trainers' Attentional States." *International Journal of Comparative Psychology* 23, no. 3 (2010). https://doi.org/10.46867/IJCP.2010.23.03.03.

000 **and corvids:** Van Der Vaart, Elske, Rineke Verbrugge, and Charlotte K. Hemelrijk. "Corvid Re-Caching without 'Theory of Mind': A Model." *PLoS ONE* 7, no. 3 (March 1, 2012): e32904. https://doi.org/10.1371/journal.pone.0032904.

000 **composed of simpler cognitive elements:** Urquiza-Haas, Esmeralda G., and Kurt Kotrschal. "The Mind behind Anthropomorphic Thinking: Attribution of Mental States to Other Species." *Animal Behaviour* 109 (November 2015): 167–76. https://doi.org/10.1016/j.anbehav.2015.08.011.

000 **it reflects a deep-seated impulse:** Urquiza-Haas, Esmeralda G., and Kurt Kotrschal. "The Mind behind Anthropomorphic Thinking: Attribution of Mental States to Other Species."

000 **a sense of how one ought to behave:** Andrews, Kristin. "Naïve Normativity: The Social Foundation of Moral Cognition." *Journal of the American Philosophical Association* 6, no. 1 (2020): 36–56. https://doi.org/10.1017/apa.2019.30.

000 **may quit performing:** McGetrick, Jim, and Friederike Range. "Inequity Aversion in Dogs: A Review." *Learning & Behavior* 46, no. 4 (December 2018): 479–500. https://doi.org/10.3758/s13420-018-0338-x.

000 **Crows and ravens ... a similar aversion:** Wascher, Claudia A. F., and Thomas Bugnyar. "Behavioral Responses to Inequity in Reward Distribution and Working Effort in Crows and Ravens." *PLoS ONE* 8, no. 2 (February 20, 2013): e56885. https://doi.org/10.1371/journal.pone.0056885.

000 **writes ethologist Marc Bekoff:** Dugatkin, Lee Alan, and Marc Bekoff. "Play and the Evolution of Fairness: A Game Theory Model." *Behavioural Processes* 60, no. 3 (January 2003): 209–14. https://doi.org/10.1016/S0376-6357(02)00120-1.

000 **Rats behave ... the same way:** Pellis, Sergio M., Vivien C. Pellis, Jackson R. Ham, and E. J. M. Achterberg. "The Rough-and-Tumble Play of Rats as a Natural Behavior Suitable for Studying the Social Brain." *Frontiers in Behavioral Neuroscience* 16 (October 18, 2022): 1033999. https://doi.org/10.3389/fnbeh.2022.1033999.

000 **Orca clans ... line up to greet:** Stiffler, Lisa. "Understanding Orca Culture." *Smithsonian Magazine.*

000 **Sac-winged bat mothers vocalize:** Fernandez, Ahana Aurora, and Mirjam Knörnschild. "Pup Directed Vocalizations of Adult Females and Males in a Vocal Learning Bat." *Frontiers in Ecology and Evolution* 8 (August 14, 2020): 265. https://doi.org/10.3389/fevo.2020.00265.

000 **the pups themselves babble:** Fernandez, Ahana A., Lara S. Burchardt, Martina Nagy, and Mirjam Knörnschild. "Babbling in a Vocal Learning Bat Resembles Human Infant Babbling." *Science* 373, no. 6557 (August 20, 2021): 923–26. https://doi.org/10.1126/science.abf9279.

000 **creating a 'biological market':** Carter, Gerald G., and Gerald S. Wilkinson. "Food Sharing in Vampire Bats: Reciprocal Help Predicts Donations More than Relatedness or Harassment." *Proceedings of the Royal Society B: Biological Sciences* 280, no. 1753 (February 22, 2013): 20122573. https://doi.org/10.1098/rspb.2012.2573.

000 **Fruit bat vocalizations:** Prat, Yosef, Mor Taub, and Yossi Yovel. "Everyday Bat Vocalizations Contain Information about Emitter, Addressee, Context, and Behavior." *Scientific Reports* 6, no. 1 (December 22, 2016): 39419. https://doi.org/10.1038/srep39419.

000 **New colonies ... long-term affinities:** Willis, Craig K. R., and R. Mark Brigham. "Roost Switching, Roost Sharing and Social Cohesion: Forest-Dwelling Big Brown Bats, *Eptesicus fuscus*, Conform to the Fission–Fusion Model." *Animal Behaviour* 68, no. 3 (September 2004): 495–505. https://doi.org/10.1016/j.anbehav.2003.08.028.

000 **Honeybees take days to decide:** Seeley, Thomas D., and Susannah C. Buhrman. "Group Decision Making in Swarms of Honey Bees." *Behavioral Ecology and Sociobiology* 45, no. 1 (January 14, 1999): 19–31. https://doi.org/10.1007/s002650050536.

000 **chickadees ... deliberate about when to start foraging:** Course, Christopher J. "Investigating Social Dynamics Using Automated Radiotracking of Winter Flocks of Black-Capped Chickadees (*Poecile atricapillus*)." PhD dissertation, University of Western Ontario, 2021.

000 **actively negotiating while in flight:** Santos, Carlos D., Sebastian Przybyzin, Martin Wikelski, and Dina K. N. Dechmann. "Collective Decision-Making in Homing Pigeons: Larger Flocks Take Longer to Decide but Do Not Make Better Decisions." *PLOS ONE* 11, no. 2 (February 10, 2016): e0147497. https://doi.org/10.1371/journal.pone.0147497.

000 **Red deer groups ... agrees on a direction:** Conradt, L., and T. J. Roper. "Group

Decision-Making in Animals." *Nature* 421, no. 6919 (January 2003): 155–58. https://doi.org/10.1038/nature01294.

000 **American bison evidently require a majority:** Shaw, Ryan A. "Social Organization and Decision Making in North American Bison: Implications for Management." PhD dissertation, Utah State University, 2011. https://digitalcommons.usu.edu/etd/1204/.

000 **Among hamadryas baboons:** Breuer, Georg. *Sociobiology and the Human Dimension*, rev. English ed. New York: Cambridge University Press, 1983.

000 **African wild dogs vote by sneezing:** Walker, Reena H., Andrew J. King, J. Weldon McNutt, and Neil R. Jordan. "Sneeze to Leave: African Wild Dogs (*Lycaon pictus*) Use Variable Quorum Thresholds Facilitated by Sneezes in Collective Decisions." *Proceedings of the Royal Society B: Biological Sciences* 284, no. 1862 (September 13, 2017): 20170347. https://doi.org/10.1098/rspb.2017.0347.

000 **whooper swans . . . signal their preference:** Black, Jeffrey M. "Preflight Signalling in Swans: A Mechanism for Group Cohesion and Flock Formation." *Ethology* 79, no. 2 (April 26, 2010): 143–57. https://doi.org/10.1111/j.1439-0310.1988.tb00707.x.

4. A NEW NATURE ETHIC

000 *A physiological demonstration with vivisection of a dog:* Franco, Nuno. "Animal Experiments in Biomedical Research: A Historical Perspective." *Animals* 3, no. 1 (March 19, 2013): 238–73. https://doi.org/10.3390/ani3010238.

000 **"The administered beatings . . . .":** Guerrini, Anita. *Experimenting with Humans and Animals: From Galen to Animal Rights.* Johns Hopkins Introductory Studies in the History of Science. Baltimore, MD: The Johns Hopkins University Press, 2003.

000 **Descartes famously likened animals to mechanical automata:** Huxley, Thomas. "On the Hypothesis That Animals Are Automata, and Its History." *Nature* 10, no. 253 (September 1874): 362–66. https://doi.org/10.1038/010362a0.

000 **Aristotle . . . turned his attention to animals:** Adl, Sina M., Brian S. Leander, Alastair G. B. Simpson, John M. Archibald, O. Roger. Anderson, David Bass, Samuel S. Bowser, et al. "Diversity, Nomenclature, and Taxonomy of Protists."*Systematic Biology* 56, no. 4 (August 1, 2007): 684–89. https://doi.org/10.1080/10635150701494127.

000 **in some ways he appreciated animal intelligence:** Steiner, Gary. *Anthropocentrism and Its Discontents: The Moral Status of Animals in the History of Western Philosophy.* (Pittsburgh: University of Pittsburgh Press, 2010).

000 **this continuum existed on a *scala naturae*:** Marino, Lori. "Has Scala Naturae Thinking Come Between Neuropsychology and Comparative Neuroscience?" *International Journal of Comparative Psychology* 16, no. 1 (December 31, 2003). https://doi.org/10.46867/C4901P.

000 **In his political writings, Aristotle described animals:** Meijer, Eva. *When Animals Speak: Toward an Interspecies Democracy.* Animals in Context. (New York: New York University Press, 2019).

000 **the Stoics . . . were unambiguous:** Steiner, Gary. *Anthropocentrism and Its Discontents,* 77–78.

000 **asked the Stoic philosopher Balbus:** Long, A. A., and D. N. Sedley. *The Hellenistic Philosophers.* (New York: Cambridge University Press, 1987).

000 **they too shared a disdain for animals:** Monsó, Susana, Judith Benz-Schwarzburg, and Annika Bremhorst. "Animal Morality: What It Means and Why It Matters." *The Journal of Ethics* 22, no. 3–4 (December 2018): 283–310. https://doi.org/10.1007/s10892-018-9275-3.

000 **Plutarch acknowledged:** Steiner, *Anthropocentrism and Its Discontents*, 98.

000 **Porphyry denounced:** Preece, Rod, ed. *Awe for the Tiger, Love for the Lamb: A Chronicle of Sensibility to Animals.* (New York: Routledge, 2002).

000 **adopted . . . by medieval Christian thinkers:** Wise, Steven M. "How Nonhuman Animals Were Trapped in a Nonexistent Universe." *Animal Law Review.* 1 (1995): 15.

000 **carried on the tradition of emphasizing differences:** Salisbury, Joyce E. *The Beast Within: Animals in the Middle Ages,* 2nd ed. (New York: Routledge, 2012).

000 **Albert the Great asserted:** Salisbury, Joyce E. *The Beast Within: Animals in the Middle Ages,* 4.

000 **St. Thomas Aquinas . . . wrote:** Salisbury, Joyce E. *The Beast Within: Animals in the Middle Ages,* 4.

000 **wrote Francis Bacon:** Montuschi, Eleonora. "Order of Man, Order of Nature: Francis Bacon's Idea of 'Dominion' over Nature." Order: God's, Man's and Nature's: Discussion Paper. Centre for Philosophy of Natural and Social Science, London School of Economics and Political Science (2010). http://eprints.lse.ac.uk/60107/.

000 **Descartes presented animals' lack of language:** Guerrini, Anita. "The rhetorics of animal rights." In *Applied Ethics in Animal Research,* John Gluck, F. Barbara Orlans and Tony Di Pasquale, eds. (West Lafayette, IN: Purdue University Press, 2001), 55–76.

000 **Thomas Hobbes . . . John Locke . . . Immanuel Kant:** Monsó, "Animal Morality: What It Means and Why It Matters."

000 **as for animals, argued Kant:** Korsgaard, Christine M. *Fellow Creatures: Our Obligations to the Other Animals.* Uehiro Series in Practical Ethics. (Oxford: Oxford University Press, 2018).

000 **lamented . . . Arthur Schopenhauer:** Midgley, Mary. *Animals and Why They Matter.* (Athens, GA: University of Georgia Press, 1983).

000 **Jeremy Bentham:** Midgley, *Animals and Why They Matter,* 59.

000 **Notable Enlightenment animal sympathizers:** Midgley, *Animals and Why They Matter,* 11–13.

000 **who considered it nonsensical:** Midgley, *Animals and Why They Matter,* 140.

000 **Frédéric Ducarme and Denis Couvet write:** Ducarme, Frédéric, and Denis Couvet. "What Does 'Nature' Mean?" *Palgrave Communications* 6, no. 1 (January 31, 2020): 14. https://doi.org/10.1057/s41599-020-0390-y.

000 **a sense of animals as sharing basic mental properties with humans:** Hornborg, Anne-Christine. *Mi'kmaq Landscapes: From Animism to Sacred Ecology.* (Oxford, UK: Taylor & Francis, 2016).

000 **wrote of a morning spent with a farmer:** Audubon, John James. *Ornithological Biography, Vol III.* (Edinburgh: Adam & Charles Black, 1835), 339–41.

000 **"our fellow brethren . . . ":** Darwin, Charles. *Charles Darwin's Notebooks, 1836–1844: Geology, Transmutation of Species, Metaphysical Enquiries.* Paul H. Barrett and Peter J. Gautrey, eds. (Cambridge, UK: Cambridge University Press, 1987).

000 **He followed the implications of evolution:** Darwin, Charles. *The Descent of Man, and Selection in Relation to Sex.* (New York: D. Appleton and Company, 1872).

000 **a second-hand just-so story:** Romanes, George. *Animal Intelligence.* (New York: D. Appleton, 1883).

000 **A case can be made that Romanes:** Pence, David Evan. "How Comparative Psychology Lost Its Soul: Psychical Research and the New Science of Animal Behavior." *Studies in History and Philosophy of Science Part C: Studies in History and Philosophy of Biological and Biomedical Sciences* 82 (August 2020): 101275. https://doi.org/10.1016/j.shpsc.2020.101275.

000 **"In no case . . . evolution and development":** Morgan, C. Lloyd. *An Introduction to Comparative Psychology.* (London: Walter Scott, Limited, 1894).

000 **"conscious in the sense of . . . .":** Skinner, B.F. *About Behaviorism.* (New York: Knopf, 1974).

000 **it was considered a radical statement:** Yoon, Carol Kaesuk. "Donald R. Griffin, 88, Dies; Argued Animals Can Think." *New York Times,* November 14, 2003.

000 **something similar happened to popular naturalists:** Lutts, Ralph H. *The Nature Fakers: Wildlife, Science & Sentiment.* (Charlottesville, VA: University Press of Virginia, 1990).

000 **"animals are creatures with wants and feelings":** Lutts, *The Nature Fakers,* 147.

000 **Long, who described animals:** Lutts, *The Nature Fakers,* 94.

000 **publicly denounced them as "nature fakers":** Lutts, *The Nature Fakers,* 129.

000 **"Nature lovers were empathizing . . . .":** Lutts, *The Nature Fakers,* ix.

000 **"Machines in Fur and Feathers":** Burroughs, John. "Machines in Fur and Feathers." *The Independent,* Volume LXIV (January-June 1908): 570–574.

000 **killed no fewer than 11,400 animals:** "Roosevelt African Expedition Collects for SI." Smithsonian Institution Archives, May 1909.

000 **"every time he gets near . . . ":** Lutts, *The Nature Fakers,* 113.

000 **who described the "humanity":** Badè, William Frederic. *The Life and Letters of John Muir: Volume I.* (Boston: Houghton Mifflin Company, 1923).

000 **He took this attitude into his beloved wilderness:** "Muir Among the Animals (Collected Writings 1874–1916)." *Relevant Obscurity,* May 24, 2016.

000 **he reproached Roosevelt:** Tobias, Ronald B. *Film and the American Moral Vision of Nature: Theodore Roosevelt to Walt Disney.* (East Lansing, MI: Michigan State University Press, 2011).

000 **Singer's 1975 *Animal Liberation*:** Singer, Peter. *Animal Liberation: A New Ethics for Our Treatment of Animals.* (London: Jonathan Cape, 1975).

000 **Regan's *The Case for Animal Rights*:** Regan, Tom. *The Case for Animal Rights.* (London: Routledge and Kegan Paul, 1984).

000 **Regan and Singer had their blind spots:** Hutchins, Michael and Christian Wemmer. "Wildlife Conservation and Animal Rights: Are They Compatible?" In *Advances in Animal Welfare 1986/87,* Michael W. Fox and Linda D. Mickley, eds. (Dordrecht: Springer Netherlands, 1987), 111–137.

000 **More broad-minded was . . . Mary Midgley:** McElwain, Gregory S. "The Mixed Community." In *Science and the Self: Animals, Evolution, and Ethics: Essays in Honour of Mary Midgley,* Ian James Kidd and Liz McKinnell, eds. (London: Routledge, 2016), 41–51.

000 **Midgley was fascinated:** Crace, John. "Mary Midgley: Moral Missionary." *The Guardian*, September 20, 2005.

000 **"inevitably gives us a sense of fellowship with them":** Midgley, *Animals and Why They Matter*, 14.

000 **"Love, like compassion":** Midgley, *Animals and Why They Matter*, 119.

000 **"the animal turn":** Ritvo, Harriet. "On the Animal Turn." *Daedalus* 136, no. 4 (October 2007): 118–22. https://doi.org/10.1162/daed.2007.136.4.118.

000 **"What Is Conservation Biology?":** Soulé, Michael E. "What Is Conservation Biology?" *BioScience* 35, no. 11 (December 1985): 727–34. https://doi.org/10.2307/1310054.

000 **ideas that have rippled through conservation and environmentalism:** Kareiva, Peter, Robert Lalasz, and Michelle Marvier. "Conservation in the Anthropocene: Beyond Solitude and Fragility." *Breakthrough Journal* 2, Fall (2011): 29–37.

000 **In precolonial lore:** Holmgren, Virginia C. *Raccoons: In Folklore, History & Today's Backyards.* (Santa Barbara, CA: Capra Press, 1990).

000 **a moment of pop-culture acclaim:** Pettit, Michael. "The Problem of Raccoon Intelligence in Behaviourist America." *The British Journal for the History of Science* 43, no. 3 (September 2010): 391–421. https://doi.org/10.1017/S0007087409990677.

000 **subjects of study:** Pettit, "The Problem of Raccoon Intelligence in Behaviourist America."

000 **"Thinking Like a Mountain":** Leopold, Aldo. *A Sand County Almanac, and Sketches Here and There.* (New York: Oxford University Press, 1949).

000 **Leopold wrote about encountering raccoons:** Leopold, Aldo. *Round River: From the Journals of Aldo Leopold.* (New York: Oxford University Press, 1953).

000 **earning the condemnation of Rachel Carson:** Britt, Cynthia J. "Rachel Carson and Nature as Resource, Object, and Spirit: Identification, Consubstantiality, and Multiple Stakeholders in the Environmental Rhetoric of the *Conservation in Action* Series." Doctoral dissertation, University of Louisville, 2010. https://ir.library.louisville.edu/etd/154/.

000 **"men are only fellow-voyagers . . . ":** Leopold, *A Sand County Almanac, and Sketches Here and There*, 102.

000 **"seed stock":** Leopold, Aldo. *A Sand County Almanac, and Sketches Here and There*, 69.

000 **in the population-level and ecosystem-level terms:** Callicott, J. Baird, Jonathan Parker, Jordan Batson, Nathan Bell, Keith Brown, and Samantha Moss. "The Other in A Sand County Almanac: Aldo Leopold's Animals and His Wild-Animal Ethic." *Environmental Ethics* 33, no. 2 (2011): 115–46. https://doi.org/10.5840/enviroethics201133217.

000 **"A thing is right . . . ":** Leopold, *A Sand County Almanac, and Sketches Here and There*, 211.

5. ANIMAL PERSONHOOD

000 **To Immanuel Kant:** Kant, Immanuel. "Rational Beings Alone Have Moral Worth." In *Environmental Ethics: Readings in Theory and Application,* Louis P. Pojman, Paul Pojman, and Katie McShane, eds. (Boston: Cengage Learning, 2016), 85–7.

000 **'person' in the *Oxford English Dictionary*:** Waite, Maurice, ed. *Paperback Oxford English Dictionary.* 7th ed. (Oxford: Oxford University Press, 2012).

000 **if you look to *Black's Law Dictionary*:** Garner, Bryan A., and Henry Campbell Black, eds. *Black's Law Dictionary*, 11th ed. (St. Paul, MN: Thomson Reuters, 2019).

000 **This reflects a recent change:** Lo, Spencer. "What Is A Legal Person? Law Dictionary Corrects Decades-Old Error." Nonhuman Rights Project, June 25, 2019. https://www.nonhumanrights.org/blog/legal-person-blacks-law-correction/.

000 **states considered cruelty to animals a misdemeanor:** Keim, Brandon. "An Elephant's Personhood on Trial." *Atlantic*, December 28, 2018.

000 **every state now regards animal cruelty:** Keim, "An Elephant's Personhood on Trial."

000 **they can typically be sued:** Keim, "An Elephant's Personhood on Trial."

000 **Animal Welfare Act exempts:** "USDA APHIS | Animal Welfare Act."

000 **the Humane Slaughter Act doesn't apply:** Walsh, Owen. "Does the Humane Slaughter Act really protect animals?" The Humane League, January 17, 2023..

000 **destroyed the state's largest seabird colony:** Rago, Gordon. "Virginia Paved Over Its Largest Seabird Nesting Site during the Hampton Roads Bridge-Tunnel Expansion." *The Virginian-Pilot*, January 7, 2020.

000 **earned most of its members fines:** Schatz, Bryan. "Catching a Band of Wildlife Killers." *High Country News*, 2020.

000 **lawsuit on behalf of . . . Kama:** Citizens v. New England Aquarium, 836 F. Supp. 45 (D. Mass. 1993).

000 **"A thick and impenetrable legal wall":** Wise, Steven M. *Rattling the Cage: Towards Legal Rights for Animals*. (New York: Perseus, 2000).

000 **captured from a herd in Thailand:** Tullis, Tracy. "The Bronx Zoo's Loneliest Elephant." *New York Times*, June 26, 2015.

000 **celebrated her birth with trumpeting:** "Echo: An Elephant to Remember." *PBS*, October 11, 2010.

000 **Their destination:** Lepore, Jill. "The Elephant Who Could Be a Person." *Atlantic*, November 16, 2021.

000 **Sleepy died . . . . The other elephants were sent:** Nonhuman Rights Project. "Happy."

000 **Happy and Grumpy . . . ended up:** Nonhuman Rights Project. "Happy."

000 **"a more physical elephant":** Nonhuman Rights Project. "Happy."

000 **As adults . . . many scientists:** Bradshaw, G. A., Allan N. Schore, Janine L. Brown, Joyce H. Poole, and Cynthia J. Moss. "Elephant Breakdown." *Nature* 433, no. 7028 (February 2005): 807–807. https://doi.org/10.1038/433807a.

000 **Patty and Maxine, attacked Grumpy:** Hamilton, Brad. "Happy the Elephant's Sad Life Alone at the Bronx Zoo." *New York Post*, September 30, 2012.

000 **an indoor cage:** Hamilton, "Happy the Elephant's Sad Life Alone at the Bronx Zoo."

000 **Personal injury claims:** "Learning the Basics: Personal Injury Law." AllLaw.com.

000 **as does privacy:** Post, Robert C. "The Social Foundations of Privacy: Community and Self in the Common Law Tort." *California Law Review* 77, no. 5 (1989): 957–1010. https://doi.org/10.2307/3480641.

000 **same-sex marriages:** Nicolas, Peter. "Common Law Same-Sex Marriage," *Connecticut Law Review* 43, no. 3 (2011): 931–47.

000 **a book about *Somerset v. Stewart*:** Wise, Steven M. *Though the Heavens May Fall: The Landmark Trial That Led to the End of Slavery*. (London: Pimlico, 2006).

000 **The first hearing:** In the Matter of: Nonhuman Rights Project v. James Breheny, et al., September 23, 2019, Peter Kent, https://archive.nonhumanrights.org/content/uploads/92319-Happy-oral-arguments-corrected-transcript.pdf.

000 **affidavits from primatologists who asserted:** Nonhuman Rights Project. "Tommy." See affidavits from James R. Anderson, Christophe Boesch, Jennifer M.B. Fugate, Mary Lee Jensvold, James King, Tetsuro Matsuzawa, William C. McGrew, Mathias Osvath, and Emily Sue Savage-Rumbaugh.

000 **habeas corpus simply didn't apply:** https://archive.nonhumanrights.org/content/uploads/Appellate-Decision-in-Tommy-Case-12-4-14.pdf.

000 **appeals court upheld that ruling:** https://archive.nonhumanrights.org/content/uploads/Appellate-Decision-in-Tommy-Case-12-4-14.pdf.

000 **more sympathetic but declined:** https://archive.nonhumanrights.org/content/uploads/Transcript_of_Oral_Argument-_Niagara_County_12-9-13.pdf.

000 **an appeals court upheld the decision:** http://www.nycourts.gov/courts/ad4/Clerk/Decisions/2015/01-02-15/PDF/1300.pdf.

000 **bound by the earlier . . . ruling:** https://archive.nonhumanrights.org/content/uploads/Judge-Jaffes-Decision-7-30-15.pdf.

000 **proceeded to explain:** https://archive.nonhumanrights.org/content/uploads/92319-Happy-oral-arguments-corrected-transcript.pdf, 25–61.

000 **affidavits from five researchers:** Nonhuman Rights Project. "Happy." https://www.nonhumanrights.org/client/happy/. See affidavits from Lucy Bates and Richard W. Byrne, Karen McComb, Cynthia Moss, and Joyce Poole.

000 **beginning with their brains:** https://archive.nonhumanrights.org/content/uploads/Aff.-Joyce-Poole-1.pdf.

000 **"The physical similarities . . . ":** https://archive.nonhumanrights.org/content/uploads/Joint-Affidavit-Lucy-Bates-Richard-Byrne-1.pdf, 16.

000 **experiment that established elephant self-awareness:** Plotnik, Joshua M., Frans B. M. De Waal, and Diana Reiss. "Self-Recognition in an Asian Elephant." *Proceedings of the National Academy of Sciences* 103, no. 45 (November 7, 2006): 17053–57. https://doi.org/10.1073/pnas.0608062103.

000 **With a mental self-image:** https://archive.nonhumanrights.org/content/uploads/Joint-Affidavit-Lucy-Bates-Richard-Byrne-1.pdf, 17–29.

000 **The second hearing:** https://archive.nonhumanrights.org/content/uploads/10.21-Transcript.pdf

000 **The third hearing:** https://archive.nonhumanrights.org/content/uploads/06jan2020.pdf

000 **Patty, the zoo's other elephant:** Koehl, Dan. "Patti, Asian elephant (*Elephas maximus*) located at Bronx Zoo in United States." Elephant Encyclopedia. https://www.elephant.se/database2.php?elephant_id=2445.

000 **The *New York Times* covered the death:** "Astor the Elephant Dies at 17 Months; Charmed Thousands." *New York Times*, February 1, 1983.

000 **"Some animals get in . . . .":** Nussbaum, Martha C. "Working with and for Animals: Getting the Theoretical Framework Right." *Journal of Human Development and Capabilities* 19, no. 1 (January 2, 2018): 2–18. https://doi.org/10.1080/19452829.2017.1418963.

000 **the capability approach:** Robeyns, Ingrid, and Morten Fibieger Byskov. "The Capability Approach." In *The Stanford Encyclopedia of Philosophy* (Summer 2023 edi-

tion), Edward N. Zalta and Uri Nodelman, eds. Metaphysics Research Lab, Stanford University, 2023. https://plato.stanford.edu/archives/sum2023/entries/capability-approach/.

000 **Nussbaum did submit a brief:** https://archive.nonhumanrights.org/content/uploads/IMOTheNonhumanRightsProjectIncvBreheny-amicus-ProfessorMarthaCNussbaum-AmicusBrf.pdf.

000 **rights and justice is itself problematic:** Curtin, Deane. "Toward an Ecological Ethic of Care." *Hypatia* 6, no. 1 (1991): 60–74. https://doi.org/10.1111/j.1527-2001.1991.tb00209.x.

000 **roiled the fields:** Chelsea Batavia, correspondence with the author.

000 **the words of philosopher Karen Warren:** Warren, Karen J. "The Power and the Promise of Ecological Feminism." *Environmental Ethics* 12, no. 2 (1990): 125–46. https://doi.org/10.5840/enviroethics199012221.

000 **paid trappers . . . with neck-encircling snares:** Austin, Phyllis. "IF&W Biologist Critical of Coyote Snaring Slated for Demotion." *Maine Environmental News,* March 11, 2003.

000 **an invocation widespread:** King, Thomas, editor. *All My Relations: An Anthology of Contemporary Canadian Native Fiction.* (Toronto: McClelland & Stewart, 1998), ix.

000 **an essay by Margaret Robinson:** Robinson, Margaret. "Animal Personhood in Mi'kmaq Perspective." *Societies* 4, no. 4 (December 3, 2014): 672–88. https://doi.org/10.3390/soc4040672.

000 **angering people who owned property nearby:** Lucy Quimby, interview with the author.

000 **some of them breached:** "Beaver bog attacked, pair busted." *Associated Press,* June 24, 2003.

000 **translates to "the laughing berry":** Pollard learned this from Carol Dana, a Penobscot elder and language teacher.

000 **breached the dam again:** "Beaver bog attacked, pair busted." *Associated Press,* June 24, 2003

000 **they can be trapped . . . even waived:** Maine Department of Inland Fisheries and Wildlife. *Summary of Trapping Laws 2022–2023.*

000 **for millennia cultures have used this:** Logan, William Bryant. "Woods Work." *Emergence Magazine,* January 10, 2020.

000 **with coppices containing several times:** Logan, "Woods Work."

000 **one-sixth of Maine's beavers were killed:** Hilton, Henry. "Beaver Assessment." Maine Department of Insland Fisheries and Wildlife, April 23, 1986. https://www.maine.gov/ifw/docs/beaver-speciesassessment.pdf.

000 **annihilated North American beaver populations:** Gabaree, Rob. "The History of Beavers in New England With Ben Goldfarb." *New England News Collaborative,* August 2, 2018.

000 **Edible Landscape Project:** Sarnacki, Aislinn. "Bangor Trails Will Become 'Edible Landscapes,' Abundant in Berries, Nuts and Fruit." *Bangor Daily News,* September 10, 2019.

000 **Tuitt issued her ruling:** https://archive.nonhumanrights.org/content/uploads/HappyFeb182020.pdf.

000 **a Change.org petition:** "End Happy the Elephant's 10 Years of Solitary Confinement." Change.org petition.

000 **Alexandria Ocasio-Cortez:** Goldiner, David. "Ocasio-Cortez Is Un-Happy about Plight of Bronx Zoo Elephant." *New York Daily News,* June 6, 2019.

000 **Bill DeBlasio:** Marsh, Julia. "De Blasio sympathizes with Happy the elephant, but 'doesn't know the details.'" *New York Post,* October 4, 2019.

000 **to declare that all animals are persons:** Shad, Sonia. "Indian High Court Recognizes Nonhuman Animals As Legal Entities." Nonhuman Rights Project, July 10, 2019. https://www.nonhumanrights.org/blog/punjab-haryana-animal-rights/.

000 **Wise testified to the Constitutional Court of Colombia:** Choplin, Lauren. "NhRP Addresses Highest Court in Colombia in Chucho Bear Rights Case." Nonhuman Rights Project, August 19, 2019. https://www.nonhumanrights.org/blog/chucho-supreme-court-hearing-colombia/.

000 **a habeas case in neighboring Ecuador:** "A Landmark Ruling for Animal Rights in Ecuador." Nonhuman Rights Project, March 23, 2022. https://www.nonhumanrights.org/blog/landmark-ruling-animal-rights-ecuador/.

000 **The resulting decision:** Corte Constitutional del Ecuador, Final Judgement No. 253-20-JH/22. https://animal.law.harvard.edu/wp-content/uploads/Final-Judgment-Estrellita-w-Translation-Certification.pdf.

000 **Five of seven judges upheld Tuitt's decision:** State of New York Court of Appeals. "No. 52, In the Matter of Nonhuman Rights Project, Inc., &c.,Appellant, v. James J. Breheny, &c., et al., Respondents." June 14, 2022. https://www.nycourts.gov/ctapps/Decisions/2022/Jun22/52opn22-Decision.pdf.

000 *Success Without Victory:* Lobel, Jules. *Success without Victory: Lost Legal Battles and the Long Road to Justice in America.* (New York: New York University Press, 2003).

000 **one chapter in particular:** Lobel, *Success without Victory: Lost Legal Battles and the Long Road to Justice in America,* 46–74.

000 **a paragraph that Wise underlined heavily:** Lobel, *Success without Victory: Lost Legal Battles and the Long Road to Justice in America,* 63.

6. CITIZEN ANIMAL

000 **Leslie Street Spit's 40,000 birds:** Gail Fraser, interview with the author. That number refers specifically to the population in 2018.

000 **the Lord . . . calling cormorants detestable:** Leviticus 11:17.

000 **Satan . . . disguised as a cormorant:** Wires, Linda R. *The Double-Crested Cormorant: Plight of a Feathered Pariah.* (New Haven: Yale University Press, 2014), 58.

000 **there was a Cormorant clan:** "Ojibwe." Milwaukee Public Museum. https://www.mpm.edu/content/wirp/ICW-51.

000 **"an obvious and potent sign":** Wires, *The Double-Crested Cormorant: Plight of a Feathered Pariah,* 59.

000 **commonly called "n—-r goose":** Wires, *The Double-Crested Cormorant: Plight of a Feathered Pariah,* 60.

000 **killed them by the boatload:** Wires, *The Double-Crested Cormorant: Plight of a Feathered Pariah,* 283.

000 **scientists disputed whether:** Wires, Linda R., and Francesca J. Cuthbert. "Historic Populations of the Double-Crested Cormorant (*Phalacrocorax auritus*): Implications for Conservation and Management in the 21st Century." *Waterbirds* 29, no.

1 (March 2006): 9–37. https://doi.org/10.1675/1524-4695(2006)29[9:HPOTDC]2.0.CO;2.

000 **DDT decimated their populations again:** Wires, "Historic Populations of the Double-Crested Cormorant (*Phalacrocorax auritus*): Implications for Conservation and Management in the 21st Century."

000 **just 5,000 remained:** Wires, *The Double-Crested Cormorant: Plight of a Feathered Pariah*, 86.

000 **125 pairs were estimated to survive:** Weseloh, D. V., and B. Collier. *The Rise of the Double-Crested Cormorant on the Great Lakes: Winning the War against Contaminants.* Great Lakes Fact Sheet 1995. Ottawa, Ontario: Minister of Environment, Canadian Wildlife Service, 1995.

000 **their diets:** Andrews, D. W., G. S. Fraser, and D. V. Weseloh. "Double-Crested Cormorants During the Chick-Rearing Period at a Large Colony in Southern Ontario: Analyses of Chick Diet, Feeding Rates and Foraging Directions." *Waterbirds* 35, no. sp1 (December 2012): 82–90. https://doi.org/10.1675/063.035.sp109.

000 **six nests there in 1990:** Taylor, Bernard, Dave Andrews, and Gail S. Fraser. "Double-Crested Cormorants and Urban Wilderness: Conflicts and Management." *Urban Ecosystems* 14, no. 3 (September 2011): 377–94. https://doi.org/10.1007/s11252-011-0165-8.

000 **"Whose stories come to matter":** Van Dooren, Thom, and Deborah Bird Rose. "Storied-Places in a Multispecies City." *Humanimalia* 3, no. 2 (February 12, 2012): 1–27. https://doi.org/10.52537/humanimalia.10046.

000 **the "animal turn":** Ritvo, Harriet. "On the Animal Turn." *Daedalus* 136, no. 4 (October 2007): 118–22. https://doi.org/10.1162/daed.2007.136.4.118.

000 **Aristotle defined humans . . . political animal:** Meijer, Eva. *When Animals Speak: Toward an Interspecies Democracy.* (New York: New York University Press, 2019), 3.

000 **"Other animals have languages":** Meijer, *When Animals Speak: Toward an Interspecies Democracy*, 223.

000 **Zoopolis: A Political Theory of Animal Rights:** Donaldson, Sue and Will Kymlicka. *Zoopolis: A Political Theory of Animal Rights.* (New York: Oxford University Press, 2011).

000 **widely lauded:** "World-Renowned Philosopher Earns Royal Society of Canada Award." *Queen's Gazette*, September 14, 2021.

000 **Some disability theorists critique:** Donaldson, *Zoopolis: A Political Theory of Animal Rights*, 105–8.

000 **Domestic creatures . . . full co-citizenship:** Donaldson, *Zoopolis: A Political Theory of Animal Rights*, 101–56.

000 **Wild animals . . . citizens of other nations:** Donaldson, *Zoopolis: A Political Theory of Animal Rights*, 156–210.

000 **"among the least recognized or protected":** Donaldson, *Zoopolis: A Political Theory of Animal Rights*, 211.

000 **would be considered denizens:** Donaldson, *Zoopolis: A Political Theory of Animal Rights*, 210–52.

000 **"the democracy of species":** Kimmerer, Robin Wall. *Braiding Sweetgrass: Indigenous Wisdom, Scientific Knowledge, and the Teachings of Plants.* (Minneapolis: Milkweed Editions, 2015).

000 **blamed them for killing the trees:** Scrivener, Leslie."30,000 Cormorants Destroying Lakeside Park." *Toronto Star*, May 20, 2009.

000 ***The Dismal State of the Great Lakes:*** Ludwig, James P. *The Dismal State of the Great Lakes: An Ecologist's Analysis of Why It Happened, and How to Fix the Mess We Have Made.* (Bloomington, IN: Xlibris LLC, 2013).

000 **He was on Naubinway Island:** Unless otherwise noted, all information about Ludwig and Cosmos comes from interviews with the author.

000 **a classic passage:** Beston, Henry. The *Outermost House: A Year of Life on the Great Beach of Cape Cod.* (New York: St. Martin's Griffin, 2006), 24.

000 **cofounded in 1990:** "History & Victories." Animal Alliance of Canada.

000 **In 2005 she also helped found:** "Our Team." Animal Protection Party of Canada.

000 **40,000 cormorants were being killed:** McDonald, Karen, Ralph Toninger, Andrea Chreston, Ilona R. Feldmann, and Gail S. Fraser. "Living with Double-Crested Cormorants *(Phalacrocorax auritus)*: A Spatial Approach for Non-Lethal Management in Toronto, Canada." *Waterbirds* 41, no. 2 (June 2018): 208–20. https://doi .org/10.1675/063.041.0215.

000 **32,000 . . . in the Great Lakes region:** McDonald, "Living with Double-Crested Cormorants *(Phalacrocorax auritus)* : A Spatial Approach for Non-Lethal Management in Toronto, Canada."

000 **fall apart when challenged in court:** Sargent, Susan. "Court Ends Cormorant Slaughter in All Eastern States." Public Employees for Environmental Responsibility, May 26, 2016.

000 **threatened a lawsuit and protests:** Liz White, interview with the author.

000 **a management group:** Chreston, Andrea. "Tommy Thompson Park Double-crested Cormorant Management Report 2018." Toronto and Region Conservation Authority, 2018. https://tommythompsonpark.ca/app/uploads/2022/06/2018-10 -31_TTPDCCO_ManagementReport2018.pdf.

000 **Ultimately the parties compromised:** McDonald, "Living with Double-Crested Cormorants *(Phalacrocorax auritus)*: A Spatial Approach for Non-Lethal Management in Toronto, Canada."

000 **Over the next eight years:** McDonald, "Living with Double-Crested Cormorants *(Phalacrocorax auritus)*: A Spatial Approach for Non-Lethal Management in Toronto, Canada."

000 **"allowed for the sustained existence":** McDonald, "Living with Double-Crested Cormorants *(Phalacrocorax Auritus)*: A Spatial Approach for Non-Lethal Management in Toronto, Canada."

000 **0.52 percent of the vote:** "By-elections March 17, 2008 – Official Voting Results." Elections Canada.

000 **3.8 percent of the . . . vote:** "Dutch Party for the Animals Gains in Parliamentary Elections: 6 Seats!" Party for the Animals.

000 **People-Animals-Nature party won:** "Portugal PM rules out coalition government after October election." Reuters. August 29, 2019. Previous election results: XIVe législature (élections du 6 octobre 2019).

000

000 **VFAR's legislative victories:** Voters for Animal Rights. "Victories." https://vfar .org/victories/.

000 **the role started in 2015:** Cronin, Melissa. "NYC Mayor Hires First-Ever Animal Welfare Champion." *The Dodo,* January 27, 2015.

000 **first position of its kind:** Sanders, Anna. "New York to Become First City in US with Office Dedicated to Animal Welfare." *New York Daily News,* October 26, 2019.

000 **instead it collaborates:** Rachel Atcheson, interview with the author.

000 **to the accomplishments:** Rachel Atcheson and Christine Kim, interviews with the author.

000 **Vink . . . suggested other political roles:** Vink, Janneke. "The Possibilities for Animal Protection in Liberal Democracies." Abstract, Animal Agency: Language, Politics, Culture, May 13, 2016, https://animalagency.files.wordpress.com/2016/05/animal-agency-amsterdam-abstracts-complete.pdf.

000 **an animal ambassador:** Sacirbey, Muhamed. "Do Animals Need a UN Ambassador?" *Huffington Post,* July 4, 2014.

000 **representatives of whales and dolphins:** Bridgeman, Laura. "What Whales Have to Teach Humans About Capitalism." *Common Dreams,* October 22, 2017. https://www.commondreams.org/views/2017/10/22/what-whales-have-teach-humans-about-capitalism.

000 **Karen Bradshaw envisions:** Bradshaw, Karen. *Wildlife as Property Owners: A New Conception of Animal Rights.* (Chicago: University of Chicago Press, 2020).

000 **"animal clans were highly respected":** Simpson, Leanne. "Looking after Gdoo-Naaganinaa: Precolonial Nishnaabeg Diplomatic and Treaty Relationships." *Wicazo Sa Review* 23, no. 2 (2008): 29–42. https://doi.org/10.1353/wic.0.0001.

000 **Ontario SwiftWatch:** Gail Fraser, interview with the author.

000 **the city's wildlife:** "Oakville Wildlife Strategy (OWLS): Wildlife Conflict Protocols." https://www.oakville.ca/getmedia/75231e17-c982-4c56-9274-4fc948fb6301/wildlife-conflict-protocols.pdf.

000 **and biodiversity:** "Oakville Strategy for Biodiversity." August 2018. https://www.oakville.ca/getmedia/a9bfbb8b-1845-4939-b3fa-782dbcab81a6/environment-biodiversity-strategy.pdf.

000 **a cautionary tale:** Gluck, John P. *Voracious Science & Vulnerable Animals: A Primate Scientist's Ethical Journey.* (Chicago: University of Chicago Press, 2016).

000 **the US Fish and Wildlife Service announced:** "US Fish and Wildlife Service Finalizes New Special Permit for Cormorant Management in Lower 48 States." US Fish & Wildlife Service, February 19, 2022.

000 **merely for having an impact:** "Migratory Bird Permits; Management of Conflicts Associated With Double- Crested Cormorants (*Phalacrocorax auritus*) Throughout the United States." *Federal Register* 85, no. 249 (December 29, 2020): 85535–85556.

000 **a comment submitted on the plan:** Linda Wires, written comment shared with author.

000 **and yet it's hard to know:** Ovegård, Maria K., Niels Jepsen, Mikaela Bergenius Nord, and Erik Petersson. "Cormorant Predation Effects on Fish Populations: A Global Meta-analysis." *Fish and Fisheries* 22, no. 3 (May 2021): 605–22. https://doi.org/10.1111/faf.12540.

000 **not a single representative:** Seng, Phil. "Double-crested Cormorants and Free-Swimming Fish: Regional Information-Gathering Meetings Meeting Summary." US Fish and Wildlife Service, December 17, 2018.

000 **article in the journal *Waterbirds*:** Batavia, Chelsea, and Michael Paul Nelson. "Ethical Foundations for the Lethal Management of Double-Crested Cormorants (*Phalocrocorax auritus*) in the Eastern United States: An Argument Analysis." *Waterbirds* 41, no. 2 (June 2018): 198–207. https://doi.org/10.1675/063.041.0214.

000 **an essentially unlimited cormorant-shooting season:** Nanowski, Natalie. "Not Many People Like Cormorants, but Should Hunters Be Allowed to Kill 50 Birds per Day?" *CBC News*, December 6, 2018.

000 **an outcry . . . from conservationists:** Paas-Lang, Christian. "Ontario Government Proposes Hunting Season for Cormorants." *Globe and Mail*, December 14, 2018.

000 **and government scientists:** Pfeffer, Amanda. "Government Scientists Warn about Safety, Impacts of Proposed Cormorant Hunt." *CBC News*, March 15, 2019.

000 **the plan was reconfigured:** "Proposal to Establish a Hunting Season for Double-Crested Cormorants in Ontario." Environmental Registry of Ontario, July 31, 2020.

7. REDEEMING THE PEST

000 **Gates was the kid who would:** Unless otherwise noted, information about Gates comes from interviews with the author.

000 **most every large conservation organization:** Luther, Erin. "Urban Wildlife Organizations and the Institutional Entanglements of Conservation's Urban Turn." *Society & Animals* 26, no. 2 (April 10, 2018): 186–96. https://doi.org/10.1163/15685306-12341587.

000 **"For those who feel . . . .":** Hadidian, John. "Wildlife in US Cities: Managing Unwanted Animals." *Animals* 5, no. 4 (November 11, 2015): 1092–1113. https://www.ncbi.nlm.nih.gov/pmc/articles/PMC4693205/.

000 **almost any animal can be killed:** "Harass, Capture or Kill a Wild Animal Damaging Private Property." Government of Ontario. Accessed August 26, 2023.

000 **it's enough to be judged a nuisance:** New York Environmental Conservation Law §11-0523.

000 **beloved icons to messy encumbrance:** Locke, Charley. "The Alaskan Harbor Where Bald Eagles Scavenge Like Pigeons." *Wired*, June 15, 2017.

000 **a program they ran:** Pacelle, Wayne. "What's Possible for Possums, Best for Beavers, and Good for Gophers." *A Humane World*, September 19, 2013. https://blog.humanesociety.org/2013/09/humane-wildlife-services.html.

000 **a more humane option:** "Live-Trapping Nuisance Raccoons Means Killing Them." For Fox Sake Wildlife Rescue, March 15, 2019.

000 **Indiana tried unsuccessfully:** Bowman, Sarah. "DNR Director: 'We Heard You,' Says No to Bobcat Hunting and Killing Nuisance Wildlife." *Indianapolis Star*, May 15, 2018.

000 **1,500 raccoons yearly:** Hadidian, "Wildlife in US Cities: Managing Unwanted Animals."

000 **requires . . . raccoons to be killed:** "Raccoons." WildlifeNYC. https://www.nyc.gov/site/wildlifenyc/animals/raccoons.page.

000 **DC passed a law in 2010:** "Wildlife Control in the District." District of Columbia Department of Energy & Environment. https://doee.dc.gov/service/wildlife-control-district.

000 **cultural DNA:** Hadidian, John. "Taking the "Pest" Out of Pest Control: Humaneness and Wildlife Damage Management." *Attitudes Towards Animals Collection,* 14. https://www.wellbeingintlstudiesrepository.org/acwp_sata/14

000 **mutualism rather than domination:** Manfredo, Michael J., Tara L. Teel, and Alia M. Dietsch. "Implications of Human Value Shift and Persistence for Biodiversity Conservation: Value Shift and Conservation." *Conservation Biology* 30, no. 2 (April 2016): 287–96. https://doi.org/10.1111/cobi.12619.

000 **200 den sites:** John Hadidian, interview with the author.

000 **only half gave advice:** Baker, Sandra E., Stephanie A. Maw, Paul J. Johnson, and David W. Macdonald. "Not in My Backyard: Public Perceptions of Wildlife and 'Pest Control' in and around UK Homes, and Local Authority 'Pest Control.'" *Animals* 10, no. 2 (January 30, 2020): 222. https://doi.org/10.3390/ani10020222.

000 **a favorite of mine:** Holmgren, Virginia C. *Raccoons: In Folklore, History & Today's Backyards.* (Santa Barbara, CA: Capra Press, 1990), 36–7.

000 **$31 million developing raccoon-proof bins:** Dempsey, Amy. "Toronto Built a Better Green Bin and—Oops—Maybe a Smarter Raccoon." *Toronto Star,* August 30, 2018.

000 **raccoon-*resistant*, not raccoon-*proof*:** Dempsey, "Toronto Built a Better Green Bin and—Oops—Maybe a Smarter Raccoon."

000 **originated in southern China:** Song, Ying, Zhenjiang Lan, and Michael H. Kohn. "Mitochondrial DNA Phylogeography of the Norway Rat." *PLoS ONE* 9, no. 2 (February 28, 2014): e88425. https://doi.org/10.1371/journal.pone.0088425.

000 **Rush Limbaugh even chimed in:** "Washington, DC, Law Forces Exterminators to Capture and Relocate Rats." *The Rush Limbaugh Show,* January 16, 2012.

000 **"wildlife—not rats or mice":** "Councilmember Cheh Laments the State of Public Discourse." Press Release, Office of Councilmember Mary M. Cheh, Ward 3, January 19, 2012.

000 **laugh when tickled:** Gloveli, Natalie, Jean Simonnet, Wei Tang, Miguel Concha-Miranda, Eduard Maier, Anton Dvorzhak, Dietmar Schmitz, and Michael Brecht. "Play and Tickling Responses Map to the Lateral Columns of the Rat Periaqueductal Gray." *Neuron,* July 2023, S0896627323004774. https://doi.org/10.1016/j.neuron.2023.06.018.

000 **feel regret:** Steiner, Adam P., and A. David Redish. "Behavioral and Neurophysiological Correlates of Regret in Rat Decision-Making on a Neuroeconomic Task." *Nature Neuroscience* 17, no. 7 (July 2014): 995–1002. https://doi.org/10.1038/nn.3740.

000 **plan for the future:** Crystal, Jonathon D. "Remembering the Past and Planning for the Future in Rats." *Behavioural Processes* 93 (February 2013): 39–49. https://doi.org/10.1016/j.beproc.2012.11.014.

000 **share food . . . more readily:** Schneeberger, Karin, Gregory Röder, and Michael Taborsky. "The Smell of Hunger: Norway Rats Provision Social Partners Based on Odour Cues of Need." *PLOS Biology* 18, no. 3 (March 24, 2020): e3000628. https://doi.org/10.1371/journal.pbio.3000628.

000 **take care of one another's babies:** Schultz, Lori A., and Richard K. Lore. "Communal Reproductive Success in Rats (*Rattus norvegicus*): Effects of Group Composition and Prior Social Experience." *Journal of Comparative Psychology* 107, no. 2 (1993): 216–22. https://doi.org/10.1037/0735-7036.107.2.216.

000 **jump for joy:** Webb, Christine E., Peter Woodford, and Elise Huchard. "The Study That Made Rats Jump for Joy, and Then Killed Them: The Gap between Knowledge and Practice Widens When Scientists Fail to Engage with the Ethical Implications of Their Own Work." *BioEssays* 42, no. 6 (June 2020): 2000030. https://doi.org/10.1002/bies.202000030.

000 **"The Kernel of . . . .":** Garrett, Henry James. "The Kernel of Human (or Rodent) Kindness." *New York Times*, December 29, 2018.

000 **urban life may have made them:** Keim, Brandon. "The Intriguing New Science That Could Change Your Mind About Rats." *Wired*, January 28, 2015.

000 **Rats and mice do not qualify:** "Rats, Mice, and Birds." Animal Welfare Institute. https://awionline.org/content/rats-mice-birds.

000 **likened . . . dogs and wolves:** Drake, Amber L. "Wild Rats Are Not Like Pet Rats, and You Shouldn't Try to Keep Them." *Love to Know Pets*, July 24, 2023..

000 **report that they're quite similar:** Amber Prince, interview with the author.

000 **"An uneasy truce based on parallel planes of existence":** Hunold, Christian, and Maz Mazuchowski. "Human–Wildlife Coexistence in Urban Wildlife Management: Insights from Nonlethal Predator Management and Rodenticide Bans." *Animals* 10, no. 11 (October 28, 2020): 1983. https://doi.org/10.3390/ani10111983.

000 **banned from residential use:** "New California Law Protecting Animals from Super-toxic Rat Poisons Takes Effect in 2021." Center for Biological Diversity. https://biologicaldiversity.org/w/news/press-releases/new-california-law-protecting-animals-super-toxic-rat-poisons-takes-effect-2021-2020-12-29/.

000 **the common worlds approach:** Taylor, Affrica, and Veronica Pacini-Ketchabaw. *The Common Worlds of Children and Animals: Relational Ethics for Entangled Lives.* (London: Routledge, 2020).

000 **"Demanding to know why . . . .":** Nelson, Narda. "Rats, Death, and Anthropocene Relations in Urban Canadian Childhoods," in *Research Handbook on Childhoodnature*, Amy Cutter-Mackenzie-Knowles, Karen Malone, and Elisabeth Barratt Hacking, eds. (Berlin: Springer International Handbooks of Education, 2020).

8. THE CARING CITY

000 *Answering the Call of the Wild:* Luther, Erin. *Answering the Call of the Wild: A Hotline Operator's Guide to Helping People and Wildlife.* (Toronto, Canada: Toronto Wildlife Center, 2010).

000 **30,000 queries each year:** "What We Do." Toronto Wildlife Centre. https://www.torontowildlifecentre.com/what-we-do/

000 **a passage by Barbara Smuts:** Smuts, Barbara. "Encounters with Animal Minds." *Journal of Consciousness Studies*, 8, no. 5–7 (2001): 293–309.

000 **hunted them to near-extinction:** *National Audubon Society Birds of North America.* (New York: Knopf, 2021), 46.

000 **"thought of as instinct machines":** Anthes, Emily. "Coldblooded Does Not Mean Stupid." *New York Times*, November 18, 2013.

000 **one-third of . . . budget:** "City Wildlife 2022 Annual Report." City Wildlife. http://citywildlife.org/wp-content/uploads/CITY-WILDLIFE-2022-ANNUAL-REPORT.pdf.

000 **following the Teenage Mutant Ninja Turtle craze:** USGS Nonindigenous Aquatic Species Database. "Red-Eared Slider (*Trachemys scripta elegans*)".

000 **at least in some places:** Hopkins, Caroline. "Invasive Turtles Are Wreaking Havoc in New York City." *National Geographic,* February 21, 2020.

000 **study of red-eared slider vocalizations:** Zhou, Lu, Long-Hui Zhao, Handong Li, Tongliang Wang, Haitao Shi, and Jichao Wang. "Underwater Vocalizations of *Trachemys scripta elegans* and Their Differences among Sex–Age Groups." *Frontiers in Ecology and Evolution* 10 (November 14, 2022): 1022052. https://doi.org/10.3389/fevo.2022.1022052.

000 **a few times each day:** "What to Do about Wild Rabbits." The Humane Society of the United States. Accessed August 27, 2023. https://www.humanesociety.org/resources/what-do-about-wild-rabbits.

000 **on a trajectory to extinction:** Hurdle, Jon. "Red Knots in Steepest Decline in Years, Threatening the Species' Survival." *New York Times*, June 5, 2021.

000 **wrote journalist Emily Sohn:** Sohn, Emily. "What Is the Moral Worth of Rescuing Individual Wild Animals?" *Aeon*, August 31, 2015. https://aeon.co/essays/what-is-the-moral-worth-of-rescuing-individual-wild-animals.

000 **survival rates are often low:** Cope, Holly R., Clare McArthur, Christopher R. Dickman, Thomas M. Newsome, Rachael Gray, and Catherine A. Herbert. "A Systematic Review of Factors Affecting Wildlife Survival during Rehabilitation and Release." *PLOS ONE* 17, no. 3 (March 17, 2022): e0265514. https://doi.org/10.1371/journal.pone.0265514.

000 **barely 10 percent survived:** Sharp, Brian E. "Post-release Survival of Oiled, Cleaned Seabirds in North America." *Ibis* 138, no. 2 (June 28, 2008): 222–28. https://doi.org/10.1111/j.1474-919X.1996.tb04332.x.

000 **1,800 animals yearly:** Average of animal intake figures from City Wildlife annual reports in 2020, 2021 and 2022.

000 **expected to swell by 40 percent:** "COG, Area Governments Project More Than 1.5 Million Additional People, 1.1 Million Additional Jobs in DC Region by 2045." Metropolitan Washington Council of Governments. https://www.mwcog.org/about-us/newsroom/2016/3/9/cog-area-governments-project-more-than-15-million-additional-people-11-million-additional-jobs-in-dc-region-by-2045/.

000 **endangered species triage:** Wilson, Kerrie A., and Elizabeth A. Law. "Ethics of Conservation Triage." *Frontiers in Ecology and Evolution* 4 (September 27, 2016). https://doi.org/10.3389/fevo.2016.00112.

000 **It *is* possible to save them all:** Sample, Ian. "Cost of Saving Endangered Species £50bn a Year, Say Experts." *Guardian*, October 11, 2012.

000 **the "pigeon paradox":** Dunn, Robert R., Michael C. Gavin, Monica C. Sanchez, and Jennifer N. Solomon. "The Pigeon Paradox: Dependence of Global Conservation on Urban Nature." *Conservation Biology* 20, no. 6 (December 2006): 1814–16. https://doi.org/10.1111/j.1523-1739.2006.00533.x.

000 **part of an emerging infrastructure:** Luther, Erin. "Between *Bios* and *Philia* : Inside the Politics of Life-Loving Cities." *Urban Geography*, December 4, 2020, 1–18. https://doi.org/10.1080/02723638.2020.1854530.

000 **3,000 such requests annually:** "City Wildlife 2022 Annual Report."

000 **part of a nationwide movement:** "Lights Out Program." Audubon. https://www.audubon.org/lights-out-program.

000 **more than 5,200 birds:** Anne Lewis, correspondence with the author.

000 **600 million birds each year:** Loss, Scott R., Tom Will, Sara S. Loss, and Peter P. Marra. "Bird–Building Collisions in the United States: Estimates of Annual Mortality and Species Vulnerability." *The Condor* 116, no. 1 (February 1, 2014): 8–23. https://doi.org/10.1650/CONDOR-13-090.1.

000 **a law . . . use bird-friendly materials:** "Washington, DC Passes Bird-Friendly Building Act." American Bird Conservancy.

000 **involved red-bellied turtles:** Davis, Karen M., and Gordon M. Burghardt. "Long-term Retention of Visual Tasks by Two Species of Emydid Turtles, *Pseudemys nelsoni* and *Trachemys scripta*." *Journal of Comparative Psychology* 126, no. 3 (August 2012): 213–23. https://doi.org/10.1037/a0027827.

9. LIVING WITH COYOTES

000 **One April day in 2015:** Unless otherwise noted, all information about Scout, other named coyotes, and Janet Kessler come from interviews with the author and materials supplied by Kessler.

000 **extirpated . . . only recently returned:** "Coyotes in the Presidio." Presidio National Park. https://presidio.gov/about/sustainability/coyotes-in-the-presidio/.

000 **the work of Stan Gehrt:** "About the Project." Urban Coyote Research Project. https://urbancoyoteresearch.com/about-project.

000 **Until the early eighteenth century:** "North American Distribution." Urban Coyote Research Project. https://urbancoyoteresearch.com/coyote-info/north-american-distribution

000 **two recorded fatalities:** McNay, Mark E. "A Case History of Wolf-Human Encounters in Alaska and Canada." Alaska Department of Fish and Game Wildlife Technical Bulletin 13, 2002. https://digitalcommons.unl.edu/cgi/viewcontent.cgi?article=1025&context=wolfrecovery; Linnell, John D. C., Ekaterina Kovtun & Ive Rouart. "Wolf attacks on humans: an update for 2002–2020." NINA Report 1944, Norwegian Institute for Nature Research, 2021. https://www.wwf.de/fileadmin/fm-wwf/Publikationen-PDF/Deutschland/Report-Wolf-attacks-2002-2020.pdf

000 **a "sentinel species":** Alexander, Shelley M., and Dianne L. Draper. "The Rules We Make That Coyotes Break." *Contemporary Social Science* 16, no. 1 (January 1, 2021): 127–39. https://doi.org/10.1080/21582041.2019.1616108.

000 **often possess domestic dog DNA:** Monzón, J., R. Kays, and D. E. Dykhuizen. "Assessment of Coyote–Wolf–Dog Admixture Using Ancestry-Informative Diagnostic SNPs." *Molecular Ecology* 23, no. 1 (January 2014): 182–97. https://doi.org/10.1111/mec.12570.

000 **"with far less restraint . . . .":** Lopez, Barry Holstun. *Of Wolves and Men.* (New York: Scribner's, 1978), 139.

000 **settlers set wolves on fire . . . shut:** Coleman, Jon T. *Vicious: Wolves and Men in America.* (New Haven: Yale University Press, 2004).

000 **and then watched gleefully:** Audubon, John James. *Ornithological Biography, Vol III.* (Edinburgh: Adam & Charles Black, 1835), 339–41.

000 **one widely-cited estimate:** "Happy National Coyote Day." Project Coyote. https://projectcoyote.org/celebrate-national-coyote-day-with-the-project-coyote-pack/.

000 **day or night:** "Night Hunting Laws by State." PointOptics. https://www
.pointoptics.com/night-hunting-laws/.

000 **bodies piled on flatbed trailers:** "850 Coyotes and Foxes Killed in 2-Day Multi-State Wildlife Killing Contest." Wolf Patrol, January 21, 2020. https://wolfpatrol
.org/2020/01/21/850-coyotes-and-foxes-killed-in-2-day-multi-state-wildlife
-killing-contest/.

000 **released inside arenas:** Casey, Liam. "Ontario set to expand areas where dogs can learn to hunt live coyotes in penned areas." *CBC News*, May 5, 2023.

000 **chased down on snowmobiles:** Wilkinson, Todd. "How Bills To Stop Killing Coyotes With Snowmobiles Went Down In Flames." *Mountain Journal*, January 19, 2023..

000 **shot from helicopters:** Eichler, Fred. "Hunting Coyotes from Helicopters—Death from Above." *Free Range American*, March 17, 2022.

000 **surveyed landowners:** Alexander, Shelley M., and Dianne L. Draper. "Worldviews and Coexistence with Coyotes." In *Human–Wildlife Interactions*, Beatrice Frank, Jenny A. Glikman, and Silvio Marchini, eds. (Cambridge, UK: Cambridge University Press, 2019), 311–34. https://doi.org/10.1017/9781108235730.018.

000 **tend to see wild animals:** Manfredo, Michael J., Tara L. Teel, Andrew W. Don Carlos, Leeann Sullivan, Alan D. Bright, Alia M. Dietsch, Jeremy Bruskotter, and David Fulton. "The Changing Sociocultural Context of Wildlife Conservation." *Conservation Biology* 34, no. 6 (December 2020): 1549–59. https://doi.org/10.1111/cobi.13493.

000 **so far supports:** Monica Serrano, interview with the author.

000 **"Brazen coyote strolls Beaverton . . . .":** Wasserstrom, Shuly. " Watch: Brazen Coyote Strolls Beaverton, Eats Squirrel." *KOIN News*, November 29, 2016.

000 **Understand that they're:** Complete instructions for navigating coyote encounters can be read on Kessler's website, coyoteyipps.com.

000 **encourages people to shout or throw:** "How to Avoid Conflicts with Coyotes." Urban Coyote Research Project. Accessed September 1, 2023. https://urbancoyoteresearch.com/coyote-info/how-avoid-conflicts-coyotes.

000 **"lost their fear":** "What to do about coyotes." The Humane Society of the United States. https://www.humanesociety.org/resources/what-do-about-coyotes.

000 **they even entered:** Margolin, Malcolm. *The Ohlone Way: Indian Life in the San Francisco-Monterey Bay Area*. (Berkeley, CA: Heyday Books, 2003), 7–9.

000 **people and lions:** Baynes-Rock, Marcus and Elizabeth Marshall Thomas. "We Are Not Equals: Socio-Cognitive Dimensions of Lion/Human Relationships." *Animal Studies Journal* 6, no. 1 (2017): 104–128.

000 **leopards in the western Himalayas:** Dhee, Vidya Athreya, John D. C. Linnell, Shweta Shivakumar, and Sat Pal Dhiman. "The Leopard That Learnt from the Cat and Other Narratives of Carnivore–Human Coexistence in Northern India." *People and Nature* 1, no. 3 (September 2019): 376–86. https://doi.org/10.1002/pan3.10039.

000 **"freeing these species . . .":** López-Bao, José Vicente, Jeremy Bruskotter, and Guillaume Chapron. "Finding Space for Large Carnivores." *Nature Ecology & Evolution* 1, no. 5 (April 20, 2017): 0140. https://doi.org/10.1038/s41559-017-0140.

000 **"Today we are the heirs . . . ":** Margolin, *The Ohlone Way*, 11.

000 **one million years ago:** Smith, Felisa A., Emma A. Elliott Smith, Amelia Villaseñor, Catalina P. Tomé, S. Kathleen Lyons, and Seth D. Newsome. "Late Pleistocene Megafauna Extinction Leads to Missing Pieces of Ecological Space in a North

American Mammal Community." *Proceedings of the National Academy of Sciences* 119, no. 39 (September 27, 2022): e2115015119. https://doi.org/10.1073/pnas.2115015119.

000 **mesopredator release:** Ritchie, Euan G., and Christopher N. Johnson. "Predator Interactions, Mesopredator Release and Biodiversity Conservation." *Ecology Letters* 12, no. 9 (September 2009): 982–98. https://doi.org/10.1111/j.1461-0248.2009.01347.x.

000 **extinction of 63 vertebrate species:** Loss, Scott R, and Peter P Marra. "Population Impacts of Free-Ranging Domestic Cats on Mainland Vertebrates." *Frontiers in Ecology and the Environment* 15, no. 9 (November 2017): 502–9. https://doi.org/10.1002/fee.1633.

000 **Cat advocates rightly note:** Lynn, William S., Francisco Santiago-Ávila, Joann Lindenmayer, John Hadidian, Arian Wallach, and Barbara J. King. "A Moral Panic over Cats." *Conservation Biology* 33, no. 4 (August 2019): 769–76. https://doi.org/10.1111/cobi.13346.

000 **where predators are shot and trapped:** "Avian Predator Management Plan." Don Edwards San Francisco Bay National Wildlife Refuge, United States Fish and Wildlife Service. https://ecos.fws.gov/ServCat/DownloadFile/973.

000 **reduce nearby cat populations:** Kays, Roland, Robert Costello, Tavis Forrester, Megan C. Baker, Arielle W. Parsons, Elizabeth L. Kalies, George Hess, Joshua J. Millspaugh, and William McShea. "Cats Are Rare Where Coyotes Roam." *Journal of Mammalogy* 96, no. 5 (September 29, 2015): 981–87. https://doi.org/10.1093/jmammal/gyv100.

000 **increase the diversity of other species:** Ritchie, "Predator Interactions, Mesopredator Release and Biodiversity Conservation."

000 **discouraging their presence:** Gehrt, Stanley D., Evan C. Wilson, Justin L. Brown, and Chris Anchor. "Population Ecology of Free-Roaming Cats and Interference Competition by Coyotes in Urban Parks." *PLoS ONE* 8, no. 9 (September 13, 2013): e75718. https://doi.org/10.1371/journal.pone.0075718.

000 **two documented fatal coyote attacks:** Baker, Rex O., and Robert M. Timm. "Coyote attacks on humans, 1970-2015: implications for reducing the risks." *Human–Wildlife Interactions* 11, no. 2 (2017): 3.

000 **25 such dog attacks:** Patronek, Gary J., Jeffrey J. Sacks, Karen M. Delise, Donald V. Cleary, and Amy R. Marder. "Co-occurrence of Potentially Preventable Factors in 256 Dog Bite–Related Fatalities in the United States (2000–2009)." *Journal of the American Veterinary Medical Association* 243, no. 12 (December 15, 2013): 1726–36. https://doi.org/10.2460/javma.243.12.1726.

000 **something that has happened twice:** Deb Campbell, San Francisco Animal Care & Control, correspondence with the author.

000 **perhaps one could draw a line:** Murie, Adolph. *Ecology of the coyote in the Yellowstone.* (United States: US Government Printing Office, 1940), 122.

10. UNDER NEW MANAGEMENT

000 **In 1984:** Unless otherwise noted, information about Fred Koontz comes from interviews with the author.

000 **"fellow beings in a social community":** Manfredo, Michael J., Esmeralda G. Urquiza-Haas, Andrew W. Don Carlos, Jeremy T. Bruskotter, and Alia M. Dietsch.

"How Anthropomorphism Is Changing the Social Context of Modern Wildlife Conservation." *Biological Conservation* 241 (January 2020): 108297. https://doi.org/10.1016/j.biocon.2019.108297.

000 **a campaign against wildlife killing contests:** Francovich, Eli. "Washington Fish and Wildlife Commission Considers Ban on Hunting Competitions." *Spokesman-Review*, February 16, 2020.

000 **making Washington the seventh state:** Block, Kitty. "Breaking News: Washington Becomes Seventh US State to Outlaw Wildlife Killing Contests." *A Humane World: Kitty Block's Blog*, September 11, 2020. https://blog.humanesociety.org/2020/09/breaking-news-washington-becomes-seventh-u-s-state-to-outlaw-wildlife-killing-contests.html.

000 **about 60,000:** Ebersole, Rene. "How Killing Wildlife in the United States Became a Game." *National Geographic*, April 27, 2022.

000 **The current system . . . can be traced:** Technical Review Committee on the Public Trust Doctrine. "The Public Trust Doctrine: Implications for Wildlife Management and Conservation in the United States and Canada." Technical Review 10-01. The Wildlife Society, September 2010.

000 **sport hunting associations pushed:** Organ, John, Shane P. Mahoney, and Valerius Geist. "Born in the Hands of Hunters." *The Wildlife Professional* 4, no. 3 (Fall 2010): 22–27.

000 **helped define the nascent discipline:** Organ, "Born in the Hands of Hunters."

000 **guiding them was a set of principles:** Organ, "Born in the Hands of Hunters."

000 **downplay the role of nonhunters:** Nelson, Michael P., John A. Vucetich, Paul C. Paquet, and Joseph K. Bump. "An Inadequate Construct?" *The Wildlife Professional* 5, no. 2 (Summer 2011): 58–60.

000 **"the art of making land produce . . . ":** Leopold, Aldo. *Game Management.* (Madison: University of Wisconsin Press, 1986).

000 **a mere five percent:** Fred Koontz, based on figures from a discussion at the Washington Department of Fish & Wildlife's Budget and Policy Advisory Group.

000 **268 imperiled nongame species:** Washington Department of Fish & Wildlife. "Threatened and Endangered Species in Washington." Accessed September 8, 2023. https://wdfw.wa.gov/get-involved/educational-resources/endangered.

000 **overwhelmingly populated by . . . consumptive users:** "State Wildlife Commissions." Wildlife for All. https://wildlifeforall.us/resources/overview-state-wildlife-management/state-wildlife-commissions/.

000 **nearly eight times more:** US Department of the Interior, US Fish and Wildlife Service, and US Department of Commerce, US Census Bureau. *2016 National Survey of Fishing, Hunting and Wildlife-Associated Recreation.* FHW/16-NAT(RV). 2018.

000 **a report by the Wildlife Society:** Technical Review Committee on the Public Trust Doctrine, "The Public Trust Doctrine: Implications for Wildlife Management and Conservation in the United States and Canada."

000 **the other was Lorna Smith:** Washington Department of Fish & Wildlife. "Governor Inslee Appoints New Fish and Wildlife Commission Members: Fred Koontz, Lorna Smith." January 5, 2021. https://wdfw.wa.gov/newsroom/news-release/governor-inslee-appoints-new-fish-and-wildlife-commission-members-fred-koontz-lorna-smith.

000 **left unfilled:** Francovich, Eli. "For Nearly a Year, Eastern Washington Hasn't Been Fully Represented on the Fish and Wildlife Commission. When Will That Change?" *The Spokesman-Review,* December 19, 2021.

000 **Thorburn . . . published an op-ed:** Thorburn, Kim. "Kim Thorburn: Conservation should not be driven by ideology." *The Spokesman-Review,* January 24, 2021.

000 **an interdepartmental guide to conservation:** "Conservation: A Commission and Department Policy Guide." https://wdfw.wa.gov/sites/default/files/2021-09/conservation_policy2fkreview.pdf.

000 **a *Northwest Sportsman* article:** Walgamott, Andy. "WDFW Commissioners Discuss New Draft Conservation Policy." *Northwest Sportsman,* September 22, 2021.

000 **"us versus them" language:** Walgamott, "WDFW Commissioners Discuss New Draft Conservation Policy."

000 **"We are a part of nature":** Walgamott, "WDFW Commissioners Discuss New Draft Conservation Policy."

000 **used to expand conservation's vision:** Cronon, William. "The Trouble With Wilderness." *New York Times,* August 13, 1995.

000 **recommended continuing the hunt:** Washington Department of Fish & Wildlife. "2022 Spring Black Bear Rules and Regulations." https://wdfw.wa.gov/about/regulations/withdrawn/2022-spring-black-bear-rules-and-regulations.

000 **"I don't think fear . . . .":** Flatt, Courtney. " Washington pauses controversial spring bear hunt." *Northwest News Network,* November 19, 2021..

000 **"overwhelming public sentiment based . . . .":** Walgamott, Andy. "WDFW Commissioners Hear It From Hunters On Spring Bear, Blues Elk." *Northwest Sportsman,* December 3, 2021..

000 **They possess extraordinary problem-solving skills:** Keim, Brandon. "Does a Bear Think in the Woods?" *Sierra Magazine,* February 26, 2019.

000 **arguably language-like:** Keim, "Does a Bear Think in the Woods?"

000 **complex matriarchal societies:** Keim, "Does a Bear Think in the Woods?"

000 **exchange favors across years:** Keim, "Does a Bear Think in the Woods?"

000 **an *Outdoor Life* article:** Lynn, Bryan. "Washington State Lost Its Spring Bear Hunt to Political Overreach—And It's Just the Beginning." *Outdoor Life,* November 29, 2021.

000 **80 percent of Washingtonians:** Public Policy Polling. "October 17–18, 2022 Poll on Washington Attitudes Toward Fish and Wildlife." https://5609432.app.box.com/s/kutlutofnc2v5fybaq7uw0z9klzisve9

000 **The department's own study:** Barker, Eric. "Monitoring documents high mountain lion predation of Blue Mountains elk; some suggest limiting hunting." *Union-Bulletin,* December 25, 2021.

000 **"This may well be a case . . . .":** Walgamott, Andy. "Blues Elk In 'Crisis,' But Not For Some FWC Members." *Northwest Sportsman,* December 2, 2021.

000 **A tribal wildlife manager said:** Walgamott, "WDFW Commissioners Hear It From Hunters On Spring Bear, Blues Elk."

000 **"The single dumbest thing . . . .":** The Dori Monson Show. "'The Backlash Was Deafening:' Listeners Prompt State Wildlife Commissioner to Resign." December 13, 2021.

000 **"In saving one cougar . . . .":** The Dori Monson Show, "'The Backlash Was Deafening:' Listeners Prompt State Wildlife Commissioner to Resign."

000 **Hundreds of species relied, directly:** Barry, Joshua M., L. Mark Elbroch, Matthew E. Aiello-Lammens, Ronald J. Sarno, Lisa Seelye, Anna Kusler, Howard B. Quigley, and Melissa M. Grigione. "Pumas as Ecosystem Engineers: Ungulate Carcasses Support Beetle Assemblages in the Greater Yellowstone Ecosystem." *Oecologia* 189, no. 3 (March 2019): 577–86. https://doi.org/10.1007/s00442-018-4315-z.

000 **or indirectly:** Peziol, Michelle, L. Mark Elbroch, Lisa A. Shipley, R. Dave Evans, and Daniel H. Thornton. "Large Carnivore Foraging Contributes to Heterogeneity in Nutrient Cycling." *Landscape Ecology* 38, no. 6 (June 2023): 1497–1509. https://doi.org/10.1007/s10980-023-01630-0.

000 **his resignation letter:** Koontz, Fred. Letter of Resignation. December 13, 2021. https://s3-us-west-2.amazonaws.com/s3-wagtail.biolgicaldiversity.org/documents/Koontz-Resignation.pdf.

000 **halt the spring bear hunt:** Francovich, Eli. "Washington Wildlife Commission Strikes Down Recreational Spring Bear Hunt." *The Spokesman-Review*, November 18, 2022.

000 **approved $23 million:** Sweeden, Paula. "Big Wins for Biodiversity, Wolves, Forests, and Climate this Legislative Session!" *Conservation Northwest*, May 19, 2023..

000 **at least 14,000 years:** Katz, Bridget. "Found: One of the Oldest North American Settlements." *Smithsonian Magazine*, April 5, 2017.

000 **coastal peoples told:** Artelle, Kyle A., Janet Stephenson, Corey Bragg, Jessie A. Housty, William G. Housty, Merata Kawharu, and Nancy J. Turner. "Values-Led Management: The Guidance of Place-Based Values in Environmental Relationships of the Past, Present, and Future." *Ecology and Society* 23, no. 3 (2018): art35. https://doi.org/10.5751/ES-10357-230335.

000 **"We are all one . . . .":** Brown, Frank and Y. Kathy Brown. *Staying the Course, Staying Alive: Coastal First Nations Fundamental truths: Biodiversity, Stewardship and Sustainability.* (Victoria, Canada: Biodiversity BC, 2009).

000 **using those creatures came with:** Brown, *Staying the Course, Staying Alive.*

000 **greed or disrespect would be punished:** Brown, *Staying the Course, Staying Alive.*

000 **outlawed traditional fishing methods:** Atlas, William I., Natalie C. Ban, Jonathan W. Moore, Adrian M. Tuohy, Spencer Greening, Andrea J. Reid, Nicole Morven, et al. "Indigenous Systems of Management for Culturally and Ecologically Resilient Pacific Salmon (*Oncorhynchus* spp.) Fisheries." *BioScience* 71, no. 2 (February 15, 2021): 186–204. https://doi.org/10.1093/biosci/biaa144.

000 **amounted to genocide:** Lafontaine, Fannie. "How Canada Committed Genocide against Indigenous Peoples, Explained by the Lawyer Central to the Determination." *The Conversation*, June 11, 2021.

000 **close relatives and teachers:** Artelle, Kyle A., Janet Stephenson, Corey Bragg, Jessie A. Housty, William G. Housty, Merata Kawharu, and Nancy J. Turner. "Values-Led Management: The Guidance of Place-Based Values in Environmental Relationships of the Past, Present, and Future." *Ecology and Society* 23, no. 3 (2018): art35. https://doi.org/10.5751/ES-10357-230335.

000 **"requiring respect and instilling . . . .":** Artelle, K. A., M. S. Adams, H. M. Bryan, C. T. Darimont, J. ('Cúagilákv) Housty, W. G. (Dúqváisḷa) Housty, J. E. Moody, et al. "Decolonial Model of Environmental Management and Conservation: Insights from Indigenous-Led Grizzly Bear Stewardship in the Great Bear Rainforest." *Eth-*

ics, Policy & Environment 24, no. 3 (September 2, 2021): 283–323. https://doi.org/10 .1080/21550085.2021.2002624.

000 **nourish beloved berry patches:** Housty, William G., Anna Noson, Gerald W. Scoville, John Boulanger, Richard M. Jeo, Chris T. Darimont, and Christopher E. Filardi. "Grizzly Bear Monitoring by the Heiltsuk People as a Crucible for First Nation Conservation Practice." *Ecology and Society* 19, no. 2 (2014): art70. https://doi .org/10.5751/ES-06668-190270.

000 **rub spruce pitch:** Daisy Sewid-Smith, via correspondence with Nancy J. Turner.

000 **eat the rhizomes:** Charlie, Luschiim Arvid, and Nancy J. Turner. *Luschiim's Plants: Traditional Indigenous Foods, Materials and Medicines.* (Madeira Park, BC: Harbour Publishing, 2021).

000 **plant names derived from bears:** Kolosova, Valeria, Ingvar Svanberg, Raivo Kalle, Lisa Strecker, Ayşe Mine Gençler Özkan, Andrea Pieroni, Kevin Cianfaglione, et al. "The Bear in Eurasian Plant Names: Motivations and Models." *Journal of Ethnobiology and Ethnomedicine* 13, no. 1 (December 2017): 14. https://doi.org/10 .1186/s13002-016-0132-9.

000 **spoke . . . with the expectation:** Jennifer Walkus, interview with the author.

000 **A recent analysis:** Henson, Lauren H., Niko Balkenhol, Robert Gustas, Megan Adams, Jennifer Walkus, William G. Housty, Astrid V. Stronen, et al. "Convergent Geographic Patterns between Grizzly Bear Population Genetic Structure and Indigenous Language Groups in Coastal British Columbia, Canada." *Ecology and Society* 26, no. 3 (2021): art7. https://doi.org/10.5751/ES-12443-260307.

000 **their bodies left:** Artelle, "Decolonial Model of Environmental Management and Conservation: Insights from Indigenous-Led Grizzly Bear Stewardship in the Great Bear Rainforest."

000 **joined to oppose trophy hunting:** Hume, Mark. "B.C. first nations ban trophy bear hunting" *The Globe and Mail,* September 12, 2012..

000 **in 2017 the . . . government agreed:** Kennedy, Merrit. "British Columbia Will Ban Grizzly Bear Trophy Hunting." *NPR,* August 15, 2017.

000 **purchase commercial hunting rights:** Raincoast Conservation Foundation. "Why we purchase commercial trophy hunting tenures." November 23, 2021. https://www.raincoast.org/2021/11/why-we-purchase-commercial-trophy -hunting-tenures/.

000 **They wanted this knowledge:** Housty, "Grizzly Bear Monitoring by the Heiltsuk People as a Crucible for First Nation Conservation Practice."

000 **There would be no radio collars:** Housty, "Grizzly Bear Monitoring by the Heiltsuk People as a Crucible for First Nation Conservation Practice."

000 **Nor would researchers . . . removed their teeth:** Housty, "Grizzly Bear Monitoring by the Heiltsuk People as a Crucible for First Nation Conservation Practice."

000 **Where logging and bear habitat overlap:** Artelle, "Decolonial Model of Environmental Management and Conservation: Insights from Indigenous-Led Grizzly Bear Stewardship in the Great Bear Rainforest."

000 **Nuxalk Bear Safety Group:** Artelle, "Decolonial Model of Environmental Management and Conservation: Insights from Indigenous-Led Grizzly Bear Stewardship in the Great Bear Rainforest."

000 **building their own ecotourism industry:** Artelle, "Decolonial Model of Envi-

ronmental Management and Conservation: Insights from Indigenous-Led Grizzly Bear Stewardship in the Great Bear Rainforest."

000 **their town's second-largest employer:** Artelle, "Decolonial Model of Environmental Management and Conservation: Insights from Indigenous-Led Grizzly Bear Stewardship in the Great Bear Rainforest."

000 **two and six million:** Megan Adams, correspondence with the author.

000 **more than a dozen . . . canneries:** Megan Adams, correspondence with the author.

000 **closed in 1957:** Thomson, Jimmy. "Grizzlies at the Table." *Beside*, November 23, 2020.

000 **a fraction of precolonization numbers:** Megan Adams, correspondence with the author.

000 **a mere 10,000 sockeye returned:** Megan Adams, correspondence with the author.

000 **they wandered the streets:** Jennifer Walkus, interview with the author.

000 **stopped defending their cubs:** Jennifer Walkus, interview with the author.

000 **saw them as victims:** Megan Adams, interview with the author.

000 **"Speak to the bears . . . .":** Unless otherwise noted, information about Walkus comes from interviews with the author.

000 *ǹàǹakila:* Adams, Megan S., Brendan Connors, Taal Levi, Danielle Shaw, Jennifer Walkus, Scott Rogers, and Chris Darimont. "Local Values and Data Empower Culturally Guided Ecosystem-Based Fisheries Management of the Wuikinuxv Bear–Salmon–Human System." *Marine and Coastal Fisheries* 13, no. 4 (July 2021): 362–78. https://doi.org/10.1002/mcf2.10171.

000 **"They're the ones with . . . .":** Jennifer Walkus, presentation at the 2020 Canadian Animal Law Conference.

000 **He sent them Megan Adams:** Unless otherwise noted, information about Adams comes from interviews and correspondence with the author.

000 **nearly two-thirds of their diets:** Adams, Megan S., Christina N. Service, Andrew Bateman, Mathieu Bourbonnais, Kyle A. Artelle, Trisalyn Nelson, Paul C. Paquet, Taal Levi, and Chris T. Darimont. "Intrapopulation Diversity in Isotopic Niche over Landscapes: Spatial Patterns Inform Conservation of Bear–Salmon Systems." *Ecosphere* 8, no. 6 (June 2017): e01843. https://doi.org/10.1002/ecs2.1843.

000 **Next they plugged:** Adams, "Local Values and Data Empower Culturally Guided Ecosystem-Based Fisheries Management of the Wuikinuxv Bear–Salmon–Human System."

000 **such is not often the case:** Chris Darimont, interview with the author.

000 **all to arrive at a simple figure:** Adams, "Local Values and Data Empower Culturally Guided Ecosystem-Based Fisheries Management of the Wuikinuxv Bear–Salmon–Human System."

000 **45 million pounds:** Adams, "Local Values and Data Empower Culturally Guided Ecosystem-Based Fisheries Management of the Wuikinuxv Bear–Salmon–Human System.

000 **some 80 vertebrate species:** Levi, Taal, Grant V. Hilderbrand, Morgan D. Hocking, Thomas P. Quinn, Kevin S. White, Megan S. Adams, Jonathan B. Armstrong, et al. "Community Ecology and Conservation of Bear–Salmon Ecosystems." *Fron-*

tiers in Ecology and Evolution 8 (December 4, 2020): 513304. https://doi.org/10 .3389/fevo.2020.513304.

000 **unusually high densities of stomata:** Levi, "Community Ecology and Conservation of Bear–Salmon Ecosystems."

000 **especially bountiful and biodiverse:** Levi, "Community Ecology and Conservation of Bear–Salmon Ecosystems."

000 **hundreds of thousands of seeds:** Shakeri, Yasaman N., Kevin S. White, and Taal Levi. "Salmon-Supported Bears, Seed Dispersal, and Extensive Resource Subsidies to Granivores." *Ecosphere* 9, no. 6 (June 2018): e02297. https://doi.org/10.1002/ ecs2.2297.

000 **the nearest store:** Megan Adams, interview with the author.

000 **not enough terrestrial vertebrates on Earth:** Bar-On, Yinon M., Rob Phillips, and Ron Milo. "The Biomass Distribution on Earth." *Proceedings of the National Academy of Sciences* 115, no. 25 (June 19, 2018): 6506–11. https://doi.org/10.1073/ pnas.1711842115.

000 **Writing in the journal *Conservation Biology*:** Wallach, Arian D., Chelsea Batavia, Marc Bekoff, Shelley Alexander, Liv Baker, Dror Ben-Ami, Louise Boronyak, et al. "Recognizing Animal Personhood in Compassionate Conservation." *Conservation Biology* 34, no. 5 (October 2020): 1097–1106. https://doi.org/10.1111/cobi .13494.

11. THE INVADERS

000 **Erick Lundgren noticed something unusual:** Unless otherwise noted, information about Lundgren comes from interviews and correspondence with the author.

000 **fewer than 600 remaining:** Tesfai, Redae T., Norman Owen-Smith, Francesca Parrini, and Patricia D. Moehlman. "Viability of the Critically Endangered African Wild Ass (*Equus africanus*) Population on Messir Plateau (Eritrea)." *Journal of Mammalogy* 100, no. 1 (February 28, 2019): 185–91. https://doi.org/10.1093/ jmammal/gyy164.

000 **"do not play a functional role . . . .":** The Wildlife Society. "Effects of an Invasive Species: Feral Horses and Burros." Fact sheet. March 2017. https://wildlife.org/wp -content/uploads/2017/05/FactSheet-HorsesAndBurros_FINAL.pdf.

000 **the foxes had . . . been introduced:** National Park Service. "Island Fox." https:// www.nps.gov/chis/learn/nature/island-fox.htm.

000 **necessitating capture of the eagles:** Collins, Paul W., Brian C. Latta, and Gary W. Roemer. "Does the Order of Invasive Species Removal Matter? The Case of the Eagle and the Pig." *PLoS ONE* 4, no. 9 (September 14, 2009): e7005. https://doi.org/10 .1371/journal.pone.0007005.

000 **monitor lizards . . . a native species:** Weijola, Valter, Varpu Vahtera, André Koch, Andreas Schmitz, and Fred Kraus. "Taxonomy of Micronesian Monitors (Reptilia: Squamata: *Varanus*): Endemic Status of New Species Argues for Caution in Pursuing Eradication Plans." *Royal Society Open Science* 7, no. 5 (May 2020): 200092. https://doi.org/10.1098/rsos.200092.

000 **descendants of domesticated animals:** Gaunitz, Charleen, Antoine Fages, Kristian Hanghøj, Anders Albrechtsen, Naveed Khan, Mikkel Schubert, Andaine Seguin-Orlando, et al. "Ancient Genomes Revisit the Ancestry of Domestic

and Przewalski's Horses." *Science* 360, no. 6384 (April 6, 2018): 111–14. https://doi .org/10.1126/science.aao3297.

000 **a few dozen miles away:** Festa-Bianchet, Marco. *"Oreamnos Americanus,* Mountain Goat." *IUCN Red List of Threatened Species,* November 30, 2019. https://doi .org/10.2305/IUCN.UK.2020-2.RLTS.T42680A22153133.en.

000 **nonnative birds now fulfill:** Vizentin-Bugoni, Jeferson, Corey E. Tarwater, Jeffrey T. Foster, Donald R. Drake, Jason M. Gleditsch, Amy M. Hruska, J. Patrick Kelley, and Jinelle H. Sperry. "Structure, Spatial Dynamics, and Stability of Novel Seed Dispersal Mutualistic Networks in Hawai'i." *Science* 364, no. 6435 (April 5, 2019): 78–82. https://doi.org/10.1126/science.aau8751.

000 **flycatchers now depended on them:** Davis, Mark A., Matthew K. Chew, Richard J. Hobbs, Ariel E. Lugo, John J. Ewel, Geerat J. Vermeij, James H. Brown, et al. "Don't Judge Species on Their Origins." *Nature* 474, no. 7350 (June 2011): 153–54. https://doi.org/10.1038/474153a.

000 **a landmark 2011 article:** Davis, "Don't Judge Species on Their Origins."

000 **the nascent compassionate conservation movement:** Bekoff, Marc, ed. *Ignoring Nature No More: The Case for Compassionate Conservatio*n. (Chicago: The University of Chicago Press, 2013).

000 **"the moral residue of conservation":** Batavia, Chelsea, Michael Paul Nelson, and Arian D. Wallach. "The Moral Residue of Conservation." *Conservation Biology* 34, no. 5 (October 2020): 1114–21. https://doi.org/10.1111/cobi.13463.

000 **argued Batavia and Wallach:** Batavia, "The Moral Residue of Conservation."

000 **overfishing . . . starving many seabirds:** Grémillet, David, Aurore Ponchon, Michelle Paleczny, Maria-Lourdes D. Palomares, Vasiliki Karpouzi, and Daniel Pauly. "Persisting Worldwide Seabird-Fishery Competition Despite Seabird Community Decline." *Current Biology* 28, no. 24 (December 2018): 4009-4013.e2. https://doi .org/10.1016/j.cub.2018.10.051.

000 **analysis of 300 Mediterranean islands:** Ruffino, L., K. Bourgeois, E. Vidal, C. Duhem, M. Paracuellos, F. Escribano, P. Sposimo, N. Baccetti, M. Pascal, and D. Oro. "Invasive Rats and Seabirds after 2,000 Years of an Unwanted Coexistence on Mediterranean Islands." *Biological Invasions* 11, no. 7 (August 2009): 1631–51. https://doi.org/10.1007/s10530-008-9394-z.

000 **menagerie of large-bodied animals:** Mead, James I., Nicholas J. Czaplewski, and Larry D. Agenbroad. "Rancholabrean (late Pleistocene) Mammals and Localities of Arizona." *Vertebrate Paleontology of Arizona. Mesa Southwest Museum Bulletin* 11 (2005): 139–80.

000 **landscape engineers and keystone species:** Moleón, Marcos, José A. Sánchez-Zapata, José A. Donázar, Eloy Revilla, Berta Martín-López, Cayetano Gutiérrez-Cánovas, Wayne M. Getz, et al. "Rethinking Megafauna." *Proceedings of the Royal Society B: Biological Sciences* 287, no. 1922 (March 11, 2020): 20192643. https://doi .org/10.1098/rspb.2019.2643.

000 **the oldest known colony:** Vasek, Frank C. "Creosote Bush: Long-Lived Clones in the Mojave Desert." *American Journal of Botany* 67, no. 2 (February 1980): 246–55. https://doi.org/10.1002/j.1537-2197.1980.tb07648.x.

000 **one of the more biodiverse places:** Phillips, Steven J., Patricia Wentworth Comus, Mark A. Dimmitt, and Linda M. Brewer, eds. *A Natural History of the Sonoran Desert,* 2nd ed. (Oakland: University of California Press, 2015).

000 **information about identity and territory:** Klingel, Hans. "Observations on Social Organization and Behaviour of African and Asiatic Wild Asses (*Equus africanus and Equus hemionus*)." *Applied Animal Behaviour Science* 60, no. 2–3 (November 1998): 103–13. https://doi.org/10.1016/S0168-1591(98)00160-9.

000 **ecologists at the University of Ghent:** Hoffmann, Maurice, Eric Cosyns, and Indra Lamoot. "Large Herbivores in Coastal Dune Management: Do Grazers Do What They Are Supposed to Do?" In *Proceedings 'Dunes and Estuaries'*, p. 249-268. VLIZ, 2005.

000 **82 plant species:** Couvreur, Martine, Eric Cosyns, Indra Lamoot, Kris Verheyen, Maurice Hoffmann, and Martin Hermy. "Donkeys as mobile links for plant seed dispersal in coastal dune ecosystems." In *Proceedings 'Dunes and Estuaries 2005'*, 279–90.

000 **described the donkeys:** Couvreur, Martine, Bart Christiaen, Kris Verheyen, and Martin Hermy. "Large Herbivores as Mobile Links between Isolated Nature Reserves through Adhesive Seed Dispersal." *Applied Vegetation Science* 7, no. 2 (November 2004): 229–36. https://doi.org/10.1111/j.1654-109X.2004.tb00614.x.

000 **observations from central Iran:** Ghasemi, A., M. Hemami, M. Senn, J. Iravani, and J. Senn. "Seed dispersal by Persian wild ass (*Eqqus hemionus onager*) in Qatruiyeh National Park, South Central Iran." Contributed Poster, International Wild Equid Conference, 2012.

000 **contracted by two-thirds:** Pires, Mathias M., Paulo R. Guimarães, Mauro Galetti, and Pedro Jordano. "Pleistocene Megafaunal Extinctions and the Functional Loss of Long-Distance Seed-Dispersal Services." *Ecography* 41, no. 1 (January 2018): 153–63. https://doi.org/10.1111/ecog.03163.

000 **up to 90 percent:** Lopez, Steve. "Witness to the Devastation in Joshua Tree National Park." *Los Angeles Times*, August 17, 2023.

000 **their individual stories:** All biographical information about Peace Ridge's donkeys comes from Emily Carman, one of their caregivers.

000 **Ethologists reviewing . . . donkey cognition:** De Santis, Marta, Samanta Seganfreddo, Morgana Galardi, Franco Mutinelli, Simona Normando, and Laura Contalbrigo. "Donkey Behaviour and Cognition: A Literature Review." *Applied Animal Behaviour Science* 244 (November 2021): 105485. https://doi.org/10.1016/j.applanim.2021.105485.

000 **domestication . . . by northern African pastoralists:** Wimpenny, Jo. *Aesop's Animals: The Science behind the Fables.* (London: Bloomsbury Sigma, 2021), 139.

000 **became central to the economies:** Wimpenny, *Aesop's Animals: The Science behind the Fables*, 140.

000 **"have invariably enjoyed . . . .":** Bough, Jill. "The Mirror Has Two Faces: Contradictory Reflections of Donkeys in Western Literature from Lucius to Balthazar." *Animals* 1, no. 1 (December 14, 2010): 56–68. https://doi.org/10.3390/ani1010056.

000 **The Biblical tale of Balaam:** Numbers 22:21–34.

000 **offers a glimpse:** Bough, "The Mirror Has Two Faces: Contradictory Reflections of Donkeys in Western Literature from Lucius to Balthazar."

000 **Shakespeare himself:** Wimpenny, *Aesop's Animals: The Science behind the Fables*, 140.

000 **described donkeys as their closest friends:** Gibson, Abraham. "Beasts of Burden: Feral Burros and the American West." In *The Historical Animal*, Susan Nance, ed. (Syracuse, NY: Syracuse University Press, 2015), 38–53.

000 **allowed their donkeys to go free:** Gibson, "Beasts of Burden: Feral Burros and the American West."

000 **"Quiet and stupid . . . .":** Gibson, "Beasts of Burden: Feral Burros and the American West."

000 **"Some people slaughtered . . . .":** Gibson, "Beasts of Burden: Feral Burros and the American West."

000 **prospectors made one final gesture:** Gibson, "Beasts of Burden: Feral Burros and the American West."

000 **a 1953 missive:** Gibson, "Beasts of Burden: Feral Burros and the American West."

000 **"Remnants of initiation . . . .":** Seegmiller, Rick F., and Robert D. Ohmart. "Ecological Relationships of Feral Burros and Desert Bighorn Sheep." *Wildlife Monographs*, no. 78 (1981): 3–58. https://www.jstor.org/stable/3830689.

000 **song and map as one:** Lundgren, Michael, and Rebecca Solnit. *Michael Lundgren: Transfigurations.* (Santa Fe: Radius Books, 2008).

000 **something of an uncertainty:** De Santis, "Donkey Behaviour and Cognition: A Literature Review."

000 **a year-long study:** Seegmiller, "Ecological Relationships of Feral Burros and Desert Bighorn Sheep."

000 **direct proportion . . . proximity to water:** McCluney, Kevin E., and John L. Sabo. "Water Availability Directly Determines per Capita Consumption at Two Trophic Levels." *Ecology* 90, no. 6 (June 2009): 1463–69. https://doi.org/10.1890/08-1626.1.

000 **in decline across the Southwest:** Foldi, Steven E. "Disappearance of a Dominant Bosque Species: Screwbean Mesquite (*Prosopis pubescens*)." *The Southwestern Naturalist* 59, no. 3 (September 2014): 337–43. https://doi.org/10.1894/F02-JEM-03.1.

000 **compiled a list of traits:** Lundgren, Erick J., Simon D. Schowanek, John Rowan, Owen Middleton, Rasmus Ø. Pedersen, Arian D. Wallach, Daniel Ramp, Matt Davis, Christopher J. Sandom, and Jens-Christian Svenning. "Functional Traits of the World's Late Quaternary Large-Bodied Avian and Mammalian Herbivores." *Scientific Data* 8, no. 1 (January 20, 2021): 17. https://doi.org/10.1038/s41597-020-00788-5.

000 **argued that mixed-up modern herbivore assemblages:** Lundgren, Erick J., Daniel Ramp, John Rowan, Owen Middleton, Simon D. Schowanek, Oscar Sanisidro, Scott P. Carroll, et al. "Introduced Herbivores Restore Late Pleistocene Ecological Functions." *Proceedings of the National Academy of Sciences* 117, no. 14 (April 7, 2020): 7871–78. https://doi.org/10.1073/pnas.1915769117.

000 **he described the well-digging donkeys:** Lundgren, Erick J., Daniel Ramp, Juliet C. Stromberg, Jianguo Wu, Nathan C. Nieto, Martin Sluk, Karla T. Moeller, and Arian D. Wallach. "Equids Engineer Desert Water Availability." *Science* 372, no. 6541 (April 30, 2021): 491–95. https://doi.org/10.1126/science.abd6775.

000 **a letter to *Science*:** Rubin, Esther S., Dave Conrad, Andrew S. Jones, and John J. Hervert. "Feral Equids' Varied Effects on Ecosystems." *Science* 373, no. 6558 (August 27, 2021): 973–973. https://doi.org/10.1126/science.abl5863.

000 **Lundgren had studied predation, too:** Lundgren, Erick J., Daniel Ramp, Owen S. Middleton, Eamonn I. F. Wooster, Erik Kusch, Mairin Balisi, William J. Ripple, et al. "A Novel Trophic Cascade between Cougars and Feral Donkeys Shapes Desert Wetlands." *Journal of Animal Ecology* 91, no. 12 (December 2022): 2348–57. https://doi.org/10.1111/1365-2656.13766.

000 **can decrease the diversity:** Midgley, Jeremy J., Bernard W. T. Coetzee, Donovan

Tye, and Laurence M. Kruger. "Mass Sterilization of a Common Palm Species by Elephants in Kruger National Park, South Africa." *Scientific Reports* 10, no. 1 (July 16, 2020): 11719. https://doi.org/10.1038/s41598-020-68679-8.

000 **wants none to live in national parks:** Erick Lundgren, interview with the author.

000 **reduce . . . by roughly two-thirds:** Bureau of Land Management. "Management Options for a Sustainable Wild Horse and Burro Program." Report to Congress, 2018. https://www.blm.gov/sites/blm.gov/files/wildhorse_2018ReporttoCongress .pdf.

000 **sold into slaughter:** American Wild Horse Campaign. "Over 1,000 Wild Horses and Burros Sold at Slaughter Auctions in 22 Months, New Report Reveals." September 23, 2022. https://americanwildhorsecampaign.org/media/over-1000-wild -horses-and-burros-sold-slaughter-auctions-22-months-new-report-reveals.

000 **more than 200 million acres:** Western Watersheds Project. "Congressional Toolkit." Accessed September 11, 2023. https://westernwatersheds.org/congressional -toolkit/.

000 **several million cows:** Conversion from official metric of 14.1 million Animal Unit Months by Erik Molvar of the Western Watersheds Project in correspondence with the author.

000 **neglected and underfunded:** Grace Kuhn of the American Wild Horse Campaign, interview with the author.

000 **apple snails upon whom:** Cattau, Christopher E., Robert J. Fletcher Jr, Rebecca T. Kimball, Christine W. Miller, and Wiley M. Kitchens. "Rapid Morphological Change of a Top Predator with the Invasion of a Novel Prey." *Nature Ecology & Evolution* 2, no. 1 (November 27, 2017): 108–15. https://doi.org/10.1038/s41559-017 -0378-1.

000 **22 such species altogether:** Wallach, Arian D., Erick J. Lundgren, William J. Ripple, and Daniel Ramp. "Invisible megafauna." *Conservation Biology* 32, no. 4 (August 2018): 962–65. https://doi.org/10.1111/cobi.13116.

000 **"invisible megafauna":** Wallach, "Invisible megafauna."

000 **three-fifths are threatened:** Ripple, William J., Guillaume Chapron, José Vicente López-Bao, Sarah M. Durant, David W. Macdonald, Peter A. Lindsey, Elizabeth L. Bennett, et al. "Saving the World's Terrestrial Megafauna." *BioScience* 66, no. 10 (October 1, 2016): 807–12. https://doi.org/10.1093/biosci/biw092.

12. WILD HEARTS

000 **what made . . . evolution so disturbing:** Wallace, Alfred Russel. *The World of Life: A Manifestation of Creative Power, Directive Mind, and Ultimate Purpose.* (New York: Moffat, Yard, 1916), 398.

000 **"He has not learned . . . .":** Leopold, Aldo. *A Sand County Almanac, and Sketches Here and There.* (Oxford, UK: Oxford University Press, 1949), 123.

000 **a "sad good":** Lynn, William. "Sad Goods." September 1, 2006. https://www .williamlynn.net/ethos-sad-goods/.

000 **intellectual groundwork for effective altruism:** Wiblin, Robert and Keiran Harris. "Yew-Kwang Ng on ethics and how to create a much happier world." *80,000 Hours,* July 26, 2018.

000 **"the study of living things . . . .":** Ng, Yew-Kwang. "Towards Welfare Biology:

Evolutionary Economics of Animal Consciousness and Suffering." *Biology & Philosophy* 10, no. 3 (July 1995): 255–85. https://doi.org/10.1007/BF00852469.

000 **a conclusion he revisited:** Groff, Zach, and Yew-Kwang Ng. "Does Suffering Dominate Enjoyment in the Animal Kingdom? An Update to Welfare Biology." *Biology & Philosophy* 34, no. 4 (August 2019): 40. https://doi.org/10.1007/s10539 -019-9692-0.

000 **Before Horta turned to philosophy:** Unless otherwise noted, information about Horta comes from interviews with the author.

000 **among the first:** Oscar Horta, interview with the author.

000 **study of wall lizards:** Johannsen, Kyle. "Animal Rights and the Problem of r-Strategists." *Ethical Theory and Moral Practice* 20, no. 2 (April 2017): 333–45. https://doi.org/10.1007/s10677-016-9774-x.

000 **"The fact is . . . ":** Horta, Oscar. "Debunking the Idyllic View of Natural Processes: Population Dynamics and Suffering in the Wild." *Télos* 17, no. 1 (2010): 73–88.

000 **up for debate:** Horta, Oscar. *"Zoopolis*, Interventions and the State of Nature." *Law, Ethics and Philosophy* (2013): 113–25.

000 **"should in general be encouraged":** Tomasik, Brian. "Habitat Loss, Not Preservation, Generally Reduces Wild-Animal Suffering." February 6, 2017. https:// reducing-suffering.org/habitat-loss-not-preservation-generally-reduces-wild -animal-suffering/.

000 **prevent . . . humanely:** Belshaw, Christopher. "Death, Pain, and Animal Life." In *The Ethics of Killing Animals*, Tatjana Višak and Robert Singer, eds. (New York: Oxford University Press, 2016), 32–50.

000 **modern-day Noah's ark:** Moen, Ole Martin. "The Ethics of Wild Animal Suffering." *Etikk i Praksis—Nordic Journal of Applied Ethics*, no. 1 (May 9, 2016): 91–104. https://doi.org/10.5324/eip.v10i1.1972.

000 *support* **killing animals like Cecil:** MacAskill, Amanda and William MacAskill. "To Truly End Animal Suffering, the Most Ethical Choice Is to Kill Wild Predators (Especially Cecil the lion)." *Quartz*, September 9, 2015.

000 **possible to use genetic engineering:** Johannsen, Kyle. "Animal Rights and the Problem of r-Strategists." *Ethical Theory and Moral Practice* 20, no. 2 (April 2017): 333–45. https://doi.org/10.1007/s10677-016-9774-x.

000 **engineering r-selected species:** Johansen, "Animal Rights and the Problem of r-Strategists."

000 **"The whole system of nature":** Wallace, *The World of Life: A Manifestation of Creative Power, Directive Mind, and Ultimate Purpose*, 369.

000 **"the total amount of suffering per year . . . .":** Dawkins, Richard. *River Out of Eden: A Darwinian View of Life*. (New York: Basic Books, 1995), 132.

000 **Browning grew up in Australia:** Unless otherwise noted, information about Browning comes from interviews with the author.

000 **"incidental compassion":** Warwick, Clifford. "Cruel world or humane nature?" *The Ecologist*, May 20, 2019. https://theecologist.org/2019/may/20/cruel-world-or -humane-nature.

000 **reduced sentience:** Browning, Heather, and Walter Veit. "Positive Wild Animal Welfare." *Biology & Philosophy* 38, no. 2 (April 2023): 14. https://doi.org/10.1007/ s10539-023-09901-5.

000 **an intrinsically positive valence:** Ginsburg, Simona and Eva Jablonka. *The Evo-*

lution of the Sensitive Soul: Learning and the Origins of Consciousness. (Cambridge, MA: The MIT Press, 2019).

000 **chipmunks in urban areas:** Lyons, Jeremy, Gabriela Mastromonaco, Darryl B. Edwards, and Albrecht I. Schulte-Hostedde. "Fat and Happy in the City: Eastern Chipmunks in Urban Environments." *Behavioral Ecology* 28, no. 6 (November 13, 2017): 1464–71. https://doi.org/10.1093/beheco/arx109.

000 **amphibians . . . find urbanization especially difficult:** Murray, Maureen H, Cecilia A Sánchez, Daniel J Becker, Kaylee A Byers, Katherine El Worsley-Tonks, and Meggan E Craft. "City Sicker? A Meta-analysis of Wildlife Health and Urbanization." *Frontiers in Ecology and the Environment* 17, no. 10 (December 2019): 575–83. https://doi.org/10.1002/fee.2126.

000 **Seychelles warblers:** Hecht, Luke. "The Importance of Considering Age When Quantifying Wild Animals' Welfare." *Biological Reviews* 96, no. 6 (December 2021): 2602–16. https://doi.org/10.1111/brv.12769.

000 **One experiment involving . . . orangutans:** Ritvo, Sarah E., and Suzanne E. MacDonald. "Preference for Free or Forced Choice in Sumatran Orangutans (*Pongo abelii*)." *Journal of the Experimental Analysis of Behavior* 113, no. 2 (March 2020): 419–34. https://doi.org/10.1002/jeab.584.

000 **Another experiment involving zebrafish:** Graham, Courtney, Marina A.G. Von Keyserlingk, and Becca Franks. "Free-Choice Exploration Increases Affiliative Behaviour in Zebrafish." *Applied Animal Behaviour Science* 203 (June 2018): 103–10. https://doi.org/10.1016/j.applanim.2018.02.005.

000 **social connections improve physical health:** Wascher, Claudia A. F., Daniela Canestrari, and Vittorio Baglione. "Affiliative Social Relationships and Coccidian Oocyst Excretion in a Cooperatively Breeding Bird Species." *Animal Behaviour* 158 (December 2019): 121–30. https://doi.org/10.1016/j.anbehav.2019.10.009.

000 **less stressed by extreme heat:** Yusishen, Michael E., Gwangseok R. Yoon, William Bugg, Ken M. Jeffries, Suzanne Currie, and W. Gary Anderson. "Love Thy Neighbor: Social Buffering Following Exposure to an Acute Thermal Stressor in a Gregarious Fish, the Lake Sturgeon (*Acipenser fulvescens*)." *Comparative Biochemistry and Physiology Part A: Molecular & Integrative Physiology* 243 (May 2020): 110686. https://doi.org/10.1016/j.cbpa.2020.110686.

000 **Five Domains model:** Mellor, David J., Ngaio J. Beausoleil, Katherine E. Littlewood, Andrew N. McLean, Paul D. McGreevy, Bidda Jones, and Cristina Wilkins. "The 2020 Five Domains Model: Including Human–Animal Interactions in Assessments of Animal Welfare." *Animals* 10, no. 10 (October 14, 2020): 1870. https://doi.org/10.3390/ani10101870.

000 **they demonstrated . . . with wild horses:** Harvey, Andrea M., Ngaio J. Beausoleil, Daniel Ramp, and David J. Mellor. "A Ten-Stage Protocol for Assessing the Welfare of Individual Non-captive Wild Animals: Free-Roaming Horses (*Equus ferus caballus*) as an Example." *Animals* 10, no. 1 (January 16, 2020): 148. https://doi.org/10.3390/ani10010148.

000 **trying to estimate individual well-being:** Hecht, Luke B. B. "Accounting for Demography in the Assessment of Wild Animal Welfare." Preprint. *bioRxiv*, October 28, 2019. https://doi.org/10.1101/819565.

000 **research on salamander welfare:** Wild Animal Initiative. "How Environment Affects the Welfare of Salamanders across their Lifetimes: Tom Luhring and Cait-

lin Gabor." August 4, 2022. https://www.wildanimalinitiative.org/blog/grantee-sirens.

000 **farming's impacts on the well-being of wild caterpillars:** Wild Animal Initiative, June 16, 2022. "Estimating the Impacts of Farmland Management on Invertebrate Welfare: Ruth Feber and Paul Johnson." https://www.wildanimalinitiative.org/blog/grantee-caterpillars.

000 **the emotional lives of octopuses:** Wild Animal Initiative. "Investigating Sentience and Emotional States in Wild Octopuses." December 12, 2022. https://www.wildanimalinitiative.org/blog/grantee-octopus/.

000 **measure early-life stress in birds:** Wild Animal Initiative. "Using Thermal Imaging to Study Early Life Stress in Birds: Paul Jerem." September 9, 2022. https://www.wildanimalinitiative.org/blog/grantee-altricial-birds.

000 **the welfare of juvenile Murray cod:** Wild Animal Initiative. "Understanding the Links between Welfare and Wild Fish Survival to Adulthood: Raf Freire." October 11, 2022. https://www.wildanimalinitiative.org/blog/grantee-murray-cod.

000 **in the words of Thoreau:** Waldman, John R. *Running Silver: Restoring Atlantic Rivers and Their Great Fish Migrations.* (Guilford, CT: Lyons Press, 2013).

000 **"the fish that feeds all":** Werman, Marco and Dave Sherwood. "Fish Win: Maine About-Face Lets Alewives Return to Canada Border River." *The World,* July 9, 2013. https://theworld.org/stories/2013-07-09/fish-win-maine-about-face-lets-alewives-return-canada-border-river.

000 **rodent personalities . . . are found:** Mortelliti, Alessio, and Allison M. Brehm. "Environmental Heterogeneity and Population Density Affect the Functional Diversity of Personality Traits in Small Mammal Populations." *Proceedings of the Royal Society B: Biological Sciences* 287, no. 1940 (December 9, 2020): 20201713. https://doi.org/10.1098/rspb.2020.1713.

000 **will eat . . . haven't before encountered:** Mortelliti, Alessio, Ilona P. Grentzmann, Shawn Fraver, Allison M. Brehm, Samantha Calkins, and Nicholas Fisichelli. "Small Mammal Controls on the Climate-Driven Range Shift of Woody Plant Species." *Oikos* 128, no. 12 (December 2019): 1726–38. https://doi.org/10.1111/oik.06643.

000 **carry them across longer distances:** Brehm, Allison M., Alessio Mortelliti, George A. Maynard, and Joseph Zydlewski. "Land-Use Change and the Ecological Consequences of Personality in Small Mammals." *Ecology Letters* 22, no. 9 (September 2019): 1387–95. https://doi.org/10.1111/ele.13324.

000 **Shy rodents are most likely:** Brehm, "Land-use Change and the Ecological Consequences of Personality in Small Mammals."

000 **a mutualistic relationship:** Brehm, Allison M., and Alessio Mortelliti. "Small Mammal Personalities Generate Context Dependence in the Seed Dispersal Mutualism." *Proceedings of the National Academy of Sciences* 119, no. 15 (April 12, 2022): e2113870119. https://doi.org/10.1073/pnas.2113870119.

000 **predators of those species:** Brehm, "Small Mammal Personalities Generate Context Dependence in the Seed Dispersal Mutualism."

000 **are no more cautious or vigilant:** Wooster, Eamonn I F, Daniel Ramp, Erick J Lundgren, Adam J O'Neill, and Arian D Wallach. "Red Foxes Avoid Apex Predation without Increasing Fear." *Behavioral Ecology* 32, no. 5 (October 20, 2021): 895–902. https://doi.org/10.1093/beheco/arab053.

000 **a study of seals:** Hammerschlag, Neil, Michael Meÿer, Simon Mduduzi Seakamela, Steve Kirkman, Chris Fallows, and Scott Creel. "Physiological Stress Responses to Natural Variation in Predation Risk: Evidence from White Sharks and Seals." *Ecology* 98, no. 12 (December 2017): 3199–3210. https://doi.org/10.1002/ecy.2049.

000 **"It is not pain . . . .":** Lynn, "Sad Goods."

000 **rabies vaccinations for raccoons:** Bittel, Jason. "Inside the Massive Effort to Tackle One of America's Greatest Rabies Threats." *National Geographic*, September 27, 2019.

000 **suggested for alpine pikas:** Wilkening, Jennifer L., Chris Ray, Nathan Ramsay, and Kelly Klingler. "Alpine Biodiversity and Assisted Migration: The Case of the American Pika (*Ochotona princeps*)." *Biodiversity* 16, no. 4 (October 2, 2015): 224–36. https://doi.org/10.1080/14888386.2015.1112304.

000 **"The question is not . . . .":** Horta, "Debunking the idyllic view of natural processes: Population dynamics and suffering in the wild."

CONCLUSION

000 **believed they were born:** Hulme, Frederick Edward. *Natural History, Lore and Legend: Being Some Few Examples of Quaint and By-gone Beliefs Gathered in from Divers Authorities, Ancient and Mediaeval, of Varying Degrees of Reliability.* (London: G. Norman, 1895), 291.

000 **in symbiosis with algae:** Keim, Brandon. "The Salamander That Has Photosynthesis Happening Inside It." *Nautilus*, March 7, 2014.

000 **composed entirely of females:** Hoffmann, Kris. "Uni-sex-a-what-now? Maine's Oddest Amphibian." Of Pools and People. https://www.vernalpools.me/uni-sex-a-what-now-maines-oddest-amphibian/.

000 **salamanders' biomass may outweigh:** Davic, Robert D., and Hartwell H. Welsh. "On the Ecological Roles of Salamanders." *Annual Review of Ecology, Evolution, and Systematics* 35, no. 1 (December 15, 2004): 405–34. https://doi.org/10.1146/annurev.ecolsys.35.112202.130116.

000 **eclipsed . . . the migrations of wildebeest:** Baldwin, Robert F., Aram J. K. Calhoun, and Phillip G. deMaynadier. "Conservation Planning for Amphibian Species with Complex Habitat Requirements: A Case Study Using Movements and Habitat Selection of the Wood Frog *Rana sylvatica*." *Journal of Herpetology* 40, no. 4 (December 2006): 442–53. https://doi.org/10.1670/0022-1511(2006)40[442:CPFASW]2.0.CO;2.

000 **went unexpectedly viral:** Svidraitė, Julija and Laima. "This Woman Creates Beautiful Memorials for Dead Animals She Comes Across and Here Are 25 of the Most Heartbreaking Ones." *Bored Panda*, September 16, 2020.

Index

[TK]